I0789946

Semiconductor Process Integration 12

Editors:

J. Murota

Y. Cao

C. Claeys

S. Deleonibus

H. Ishii

H. Iwai

A. Mai

Y. Zhao

Sponsoring Division:

 Electronics and Photonics

Published by
The Electrochemical Society
65 South Main Street, Building D
Pennington, NJ 08534-2839, USA
tel 609 737 1902
fax 609 737 2743
www.electrochem.org

ecstransactions™

Vol. 104, No. 4

Copyright 2021 by The Electrochemical Society.
All rights reserved.

This book has been registered with Copyright Clearance Center.
For further information, please contact the Copyright Clearance Center,
Salem, Massachusetts.

Published by:

The Electrochemical Society
65 South Main Street
Pennington, New Jersey 08534-2839, USA

Telephone 609.737.1902
Fax 609.737.2743
e-mail: ecs@electrochem.org
Web: www.electrochem.org

ISSN 1938-6737 (online)
ISSN 1938-5862 (print)

ISBN 978-1-60768-925-6 (PDF)

Printed in the United States of America.

Preface

The "Twelfth Symposium on Semiconductor Process Integration" provides a forum for reviewing and discussing all aspects of process integration. The previous nine international "ULSI Process Integration" symposia were held during The Electrochemical Society Meetings in Honolulu, HI, in1999; Washington, D.C., in 2001 and 2007; Paris, France, in 2003; Québec, Canada, in 2005; Vienna, Austria, in 2009; Boston, MA, in 2011; San Francisco, CA, in 2013; and Phoenix, AZ, in 2015. After that, to expand the symposium areas, the name was changed from "ULSI Process Integration" to "Semiconductor Process Integration." The two "Semiconductor Process Integration" symposia took place during the 2017 ECS Meeting at National Harbor, MD, and in 2019 at Atlanta, GA.

This issue of *ECS Transactions* includes papers presented at the "Twelfth Symposium on Semiconductor Process Integration," held during the digital 240th ECS Meeting on October 10-14, 2021. The symposium, sponsored by the ECS Electronics and Photonics Division, was international in scope and included authors from Belgium, China, Germany, Austria, Korea, Japan, and the United States. At the symposium, 38 papers were presented.

This ECST issue contains 22 manuscripts, including 16 invited papers.

The organizers are grateful to all the authors for submitting their quality manuscripts. The invited speakers are acknowledged for making this symposium possible by sharing their perspectives and insights and putting considerable effort into preparing their manuscripts. We also want to acknowledge the contributions of the Session Chairmen in chairing and guiding the sessions. The staff of The Electrochemical Society are thanked for their support.

August 2021

J. Murota, Tohoku University, Sendai, Japan
C. Claeys, KU Leuven, Leuven, Belgium
H. Iwai, National Yang Ming Chiao Tung University, Taiwan
S. Deleonibus, CEA-LETI, Grenoble, France
A. Mai, IHP, Frankfurt (Oder), Germany
Y. Cao, Qorvo Inc., Texas, US
H. Ishii, Toyohashi University of Technology, Toyohashi, Japan
Y. Zhao, Arizona State University, Tempe, Arizona, US

***ECS Transactions*, Volume 104, Issue 4**
Semiconductor Process Integration 12

Table of Contents

Preface   *iii*

Chapter 1
Advanced Device Technology 1

(Invited) Impact of Processing Factors on the Low-Frequency Noise of Gate-All-   3
Around Silicon Vertical Nanowire FETs
  E. Simoen, A. Veloso, P. Matagne, C. Claeys

Chapter 2
Advanced Device Technology 2

(Invited) HfZrO-Based Ferroelectric Devices for Lower Power AI and Memory   17
Applications
  S. Takagi, K. Toprasertpong, K. Tahara, E. Nako, R. Nakane, Z. Wang, X. Luo,
  T. E. Lee, M. Takenaka

(Invited) Vertically Stacked Junction Devices Fabricated Using Single-Crystal   27
Graphene on SiC Substrate
  M. Nagase

(Invited) Single-Electron Manipulation in an Attofarad-Capacitor DRAM   33
  K. Nishiguchi, K. Chida, A. Fujiwara

Chapter 3
Novel Integration

(Invited) Multi-Channel AlGaN/GaN Power Rectifiers: Breakthrough Performance up   51
to 10 kV
  Y. Zhang, M. Xiao, Y. Ma, Z. Du, H. Wang, A. Xie, E. Beam, Y. Cao, K. Cheng

Chapter 4
Metrogy and Characterization

(Invited) Millisecond Annealing by Atmospheric Pressure Thermal Plasma Jet and
Direct Imaging of Temperature Distribution Using Optical Interference Contactless
Thermometry (OICT)
    S. Higashi, K. Matsuguchi, T. Sato, H. Hanafusa

63

(Invited) Two-Dimensional Characterization of Wide-Bandgap Materials and Contact
Interfaces by Using Scanning Internal Photoemission Microscopy
    K. Shiojima

69

(Invited) A Highly Sensitive MEMS Accelerometer Module for Measuring Micro
Muscular Sounds
    H. Ito, K. Machida, N. Ishihara, Y. Miyake, K. Masu

83

(Invited) Nanoscale Probing of Field-Driven Ion Migration in TaO_x for Neuromorphic
Memristor Applications
    A. Tsurumaki-Fukuchi, T. Katase, H. Ohta, M. Arita, Y. Takahashi

93

Chapter 5
Novel Materials and Characterization 1

(Invited) Impact of Boron Doping and H_2 Annealing on Light Emission from Ge/Si
Core-Shell Quantum Dots
    S. Miyazaki, K. Makihara

105

(Invited) Study of HfO_2-Based High-k Gate Insulators for GaN Power Device
    T. Nabatame, E. Maeda, M. Inoue, M. Hirose, R. Ochi, T. Sawada, Y. Irokawa,
    T. Hashizume, K. Shiozaki, T. Onaya, K. Tsukagoshi, Y. Koide

113

Importance of Annealing Step on Dielectric Constant of ZrO_2 Layer of MIM
Capacitors with Al_2O_3/ZrO_2 and ZrO_2/Al_2O_3 Stack Structures
    T. Sawada, T. Nabatame, T. Onaya, M. Inoue, A. Ohi, N. Ikeda, K. Tsukagoshi

121

Study of SiO_2 Interfacial Layer Growth during Fabrication Process of Ferroelectric
$Hf_xZr_{1-x}O_2$-Based Metal-Ferroelectric Semiconductor
    T. Onaya, T. Nabatame, M. Inoue, T. Sawada, H. Ota, Y. Morita

129

Chapter 6
Novel Process-Growth

(Invited) Cutting-Edge Epitaxial Processes for Sub 3 Nm Technology Nodes: 139
Application to Nanosheet Stacks and Epitaxial Wrap-Around Contacts
   A. Y. Hikavyy, C. Porret, M. Mencarelli, R. Loo, P. Favia, M. Ayyad, B. Briggs,
   R. Langer, N. Horiguchi

(Invited) Selective Epitaxy of Submicron Ge Wire Structures for Photodetectors and   147
Optical Modulators in Si Photonics
   Y. Ishikawa, K. Noguchi, M. Tachibana, K. Kawashita, R. Oyamada, K. Motomura,
   S. Sonoi, R. Katamawari, T. Hizawa

(Invited) Fabrication of Ge-on-Insulator By Epitaxial Growth and Ion-Implanted   157
Exfoliation for Electronics and Optoelectronics Applications
   K. Yamamoto, D. Wang, H. Nakashima

Effect of Strain on the Epitaxy of B-Doped $Si_{0.5}Ge_{0.5}$ Source/Drain Layers   167
   G. Rengo, C. Porret, A. Y. Hikavyy, E. Rosseel, M. Ayyad, R. J. H. Morris,
   G. Pourtois, R. Loo, A. Vantomme

Chapter 7
Novel Materials and Characterization 2

(Invited) Thermoelectric Properties of Tin-Incorporated Group-IV Thin Films   183
   M. Kurosawa, O. Nakatsuka

Chapter 8
Novel Process-Nanofabrication

(Invited) Study on Impact of MOS Interface Passivation Processes on Band   193
Alignment and Flat-Band Voltage of 4H-SiC Gate Stacks
   K. Kita

Modeling of Advanced FinFET Dummy Gate Corner Residue Impacted By Clogging   201
   X. Xiao, X. Ke, B. Su, H. Y. Zhang

Process Development of Dislocation-Free SiGe P-Channel in FinFET Technology 209
  B. Su, W. F. Deng, C. Yin, E. N. Zhang, X. Ke, H. Oh, J. Zhao, B. Ye, H. Y. Zhang

Highly Selective SiGe Dry Etch Process for the Enablement of Stacked Nanosheet 217
Gate-All-Around Transistors
  C. Durfee, S. Kal, S. Pancharatnam, M. Bhuiyan, I. Otto IV, M. Flaugh, J. Smith,
  D. Chanemougame, C. Alix, H. Zhou, J. Frougier, A. Greene, M. Belyansky,
  K. Watanabe, J. Zhang, D. Schmidt, M. Breton, K. Zhao, M. Wang, V. Basker,
  A. Mosden, N. Loubet, D. Guo, P. Biolsi, B. Haran, H. Bu

Author Index 229

Facts about ECS

The Electrochemical Society (ECS) is an international, nonprofit, scientific, educational organization advancing the theory and practice of electrochemistry and solid state science and technology, and allied subjects. The Society was founded in Philadelphia in 1902 and incorporated in 1930. There are currently over 8,000 members from around the globe representing 13 technical division and 23 geographical sections and a growing student membership program with over 100 student chapters. The Society is also supported by more than 2,000 corporations, government agencies, and academic institutions through institutional membership, corporate programs, and subscriptions.

The technical activities of the Society are carried on by divisions. Sections of the Society host symposia, programs, and events focused on their respective geographic regions. Major international meetings of the Society are held in the spring and fall of each year. At these meetings, the divisions and partnered organizations hold general sessions and sponsor symposia on specialized subjects.

The Society has an active publications program that includes the following:

Journal of The Electrochemical Society — (JES) is the flagship journal of The Electrochemical Society and the oldest peer-reviewed journal in its field. Since its founding in 1902, JES has evolved into one of the most highly cited and prestigious journals in electrochemistry and materials science with a cited half-life of greater than 10 years.

ECS Journal of Solid State Science and Technology — (JSS) is a peer-reviewed journal covering fundamental and applied areas of solid state science and technology, including experimental and theoretical aspects of the chemistry, and physics of materials and devices.

ECS Transactions (ECST) — is the official conference proceedings publication of The Electrochemistry Society — a high-quality venue for authors and an excellent resource for researchers. ECST offers the full-text content of proceedings from ECS meetings and ECS sponsored conferences.

The Electrochemical Society Interface — *Interface* is an authoritative yet accessible publication for those in the field of solid state and electrochemical science and technology. Published quarterly, this full-color magazine contains technical articles about the latest developments in the field, and presents news and information about the Society.

ECS Books Series — ECS books and monographs provide authoritative, detailed accounts of specific topics in electrochemistry and solid state science and technology. These titles are sponsored by ECS and published in cooperation with noted publishers such as John A. Wiley & Sons.

For more information on these publications and other Society activities, visit the ECS website:

www.electrochem.org

Chapter 1

Advanced Device Technology 1

2

Impact of Processing Factors on the Low-Frequency Noise of Gate-All-Around Silicon Vertical Nanowire FETs

E. Simoen, A. Veloso, P. Matagne and C. Claeys[a]

Imec, Kapeldreef 75, B-3001 Leuven, Belgium

[a]EE Depart, KU Leuven, Kasteelpark Arenberg 10, B-3001 Leuven, Belgium

> This paper gives an overview of the impact of certain processing steps on the low-frequency (LF) noise behavior of silicon Gate-All-Around (GAA) Vertical Nanowire (VNW) MOS transistors. It is shown that the width of a Replacement Metal Gate (RMG) cap impacts the flicker noise Power Spectral Density (PSD). The type of source contact, i.e., bulk versus a confined top contact significantly changes the nature and the magnitude of the 1/f noise. Finally, the doping density of the silicon nanowires has a subtle influence, whereby the dominant fluctuation mechanism shifts from number to mobility fluctuations, while increasing the doping density. Optimal noise performance is achieved for intermediate in-situ nanowire B doping density.

Introduction

Future CMOS technology nodes will rely on Gate-All-Around (GAA) FETs that ensure ultimate control over the short-channel effects (1)-(3). In this class of devices, GAA Vertical Nanowires (VNWs) offer some clear integration advantages over horizontal architectures. While their integration requires a somewhat more complex processing scheme (4), these devices can be implemented in a 3D architecture, for example, as selectors in memory circuits (3). When fabricating VNW transistors on a bulk substrate, the latter serves at the same time as the source, using a backside contact. It has been shown that this leads to a clear asymmetry in the DC current-voltage characteristics (5) and, as shown more recently, also in the low-frequency (LF) noise performance (6). For a true 3D implementation of VNW FETs, a separate source contact should be fabricated. This can be emulated by starting from Silicon-on-Insulator (SOI) wafers and implementing a top source contact on the silicon film.

Another processing factor that deserves some attention is the doping density of the silicon nanowires, that are generally grown by Chemical Vapor Deposition (CVD) epitaxy. Undoped silicon NWs are usually slightly n-type, while in-situ doping with B or P yields p- or n-type wires. It has been shown in the past for horizontal NWs that a junctionless p^+pp^+ or n^+nn^+ architecture may result in favorable DC (1), (2), (7) and LF noise characteristics (7)-(9). In addition, optimal noise performance has been demonstrated for medium doping density, in the range of 5×10^{18} cm^{-3} (7), (9).

LF noise is an important parameter from a viewpoint of the analog/RF application of semiconductor devices. At the same time, it reveals detailed information on the basic charge transport mechanisms and how they are impacted by the presence of processing-induced defects and imperfections. Therefore, noise has been studied in NW FETs from the early days on (10)-(29); a recent review can be found in Ref. (9). Overall, it has been demonstrated that the LF noise Power Spectral Density (PSD) can be reduced by moving from a planar, over a FinFET to a NW architecture (9), (18). At the same time, quantum confinement may give rise to novel transport and noise phenomena (26)-(28).

In this work, an overview will be given of the impact of certain processing factors on the noise spectrum, its magnitude and the underlying fluctuation mechanisms of silicon GAA VNW FETs.

Experimental

Silicon-channel NWs with a diameter of ≥26 nm (pMOS), ≥17 nm (nMOS) and a gate length (defined vertically) of up to ~100 nm have been patterned on epitaxially grown films on either 300 mm diameter bulk or Silicon-On-Insulator (SOI) substrates. The channels of these devices are either undoped (corresponding to an intrinsically undoped epitaxially Si grown layers) or in-situ B/P-doped (p/nMOS). The latter case results in a junctionless (JL) type of transistor structure. The source and drain regions of the pillars are composed of ~50 nm thick in-situ highly B-doped Si or B-SiGe for the case of the SOI-based devices. For bulk-substrate devices, on top of the pillars, additional highly B-doped SiGe (25 % Ge) was grown for obtaining an increased contact area. The gate stack consists of IL-SiO$_2$/HfO$_2$/TiN/W with an Equivalent Oxide Thickness (EOT) of ~1 nm. For the bulk transistors, a Replacement Metal Gate (RMG) scheme was used for the device fabrication, using an SiO$_2$ RMG cap, as shown schematically in Fig. 1; caps with different widths, i.e., narrow or wide, have been implemented. More process details can be found in (4).

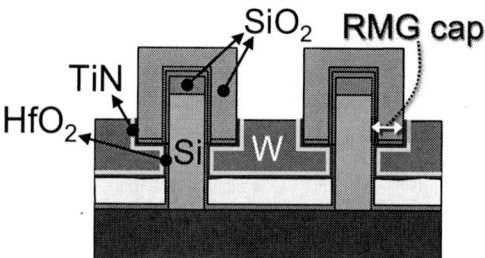

Figure 1. Schematic cross section of the vertical GAA NWFETs on bulk silicon substrate, also showing the Replacement Metal Gate (RMG) cap.

For the GAA VNW pFETs on SOI, for simplicity, a gate-first integration scheme was used with a self-aligned gate electrode process (4) and a similar gate stack as used for devices built on bulk Si substrates. A simplified schematic of these devices' configuration and an example of a Secondary Electron Microscopy (SEM) image obtained during their fabrication are shown in Fig. 2.

LF noise measurements have been performed on transistors with arrays of 10×10 NWs in parallel (= 100 NWs per device), as described elsewhere (8), (9) in linear operation (|V$_{DS}$|=0.05 mV), for a current range from about 10 nA to a few µA. Spectra have been recorded in Forward (F) or in Reverse (R) mode, with source and drain switched. The drain current noise PSD (S_I) has been evaluated at a fixed frequency f=10 Hz versus the drain current I_D; the input-referred voltage noise PSD (S_{VG}) is derived from S_I by dividing with g_m^2 (g_m the transconductance). At least 5 similar devices have been measured across the wafers to have an idea about the dispersion in the noise PSD.

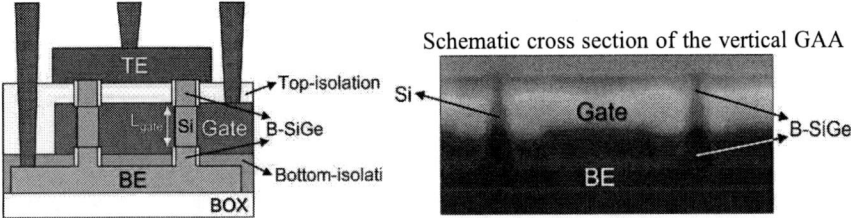

Figure 2. On the left side, a simplified schematic cross section of the vertical GAA NWFET devices built on SOI wafers. On the right side, an example of an SEM image taken during device fabrication, prior to the top electrode (TE) formation.

Results and Discussion

The impact of three processing factors on the LF noise performance of GAA VNW FETs will be discussed, namely, the width of the RMG cap, the type of source contact (bulk versus top) and the in-situ pillar B doping density for pFETs.

Impact of the RMG cap

It has been shown before that the LF noise at low frequency f is dominated by 1/f noise, irrespective of the substrate type (9). However, for the devices on bulk silicon, the spectra become white above 1 kHz (6), (9), (29). Here the focus is on the 1/f noise behavior for f<1 kHz. It has been shown before that the 1/f noise for the VNW n- and pFETs is dominated by the so-called $\Delta\mu$ fluctuations (29). This is indicated in Fig. 3 for a wide and a narrow RMG cap nMOSFET with undoped (intrinsic) silicon pillars: the S_{VG} at 10 Hz exhibits a continuous reduction with gate voltage V_{GS} up to around threshold V_T, i.e., 0.4 to 0.5 V, a fingerprint of so-called $\Delta\mu$ behavior (30), (31). This has been observed before for NW devices (11), (14), (19) and points to conduction in the core of the nanowire, away from the "noisy" traps in the gate oxide, thus suppressing the number fluctuations 1/f noise (31)-(34). The increase in S_{VG} at higher V_{GS} indicates that the dominant 1/f mechanism changes from $\Delta\mu$ to mobility fluctuations correlated with Δn (32) and further enhanced by the impact of the series resistance R_S. As a result, S_{VG} in Fig. 3 exhibits a minimum situated around V_T. In addition, both parameters are lower for the wide cap nMOSFETs (29).

To illustrate this more clearly, the V_T and the minimum S_{VG} at 10 Hz for a set of GAA VNW pFETs have been represented in Fig. 4a and 4b, respectively. The correlation between both parameters has been discussed before (29). With respect to variability of the threshold voltage, random dopant fluctuations can be ruled out, since the NWFETs in Fig. 4 have been fabricated with intrinsic, undoped pillars. This leaves process-induced variation of the EOT or the nanowire diameter (d_{NW}) as the most likely sources of the observed variability. A consistent picture emerges if it is assumed that the higher V_T and the higher 1/f noise PSD originate from a smaller average d_{NW} (over 100 NWs) for the narrow RMG cap transistors (2), (29). It also explains the higher maximum transconductance found for the wide cap VNW nFETs (29).

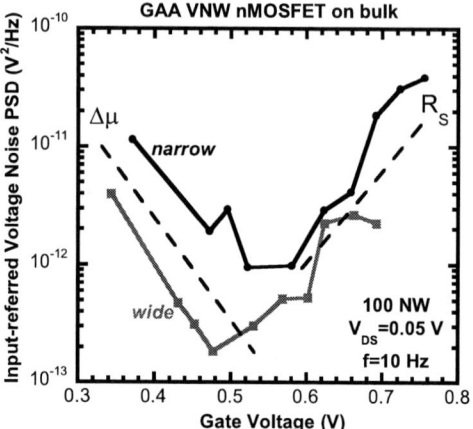

Figure 3. Input-referred noise PSD at 10 Hz for a wide and narrow cap GAA VNW nMOSFET without channel doping, in linear operation (V_{DS}=0.05 V).

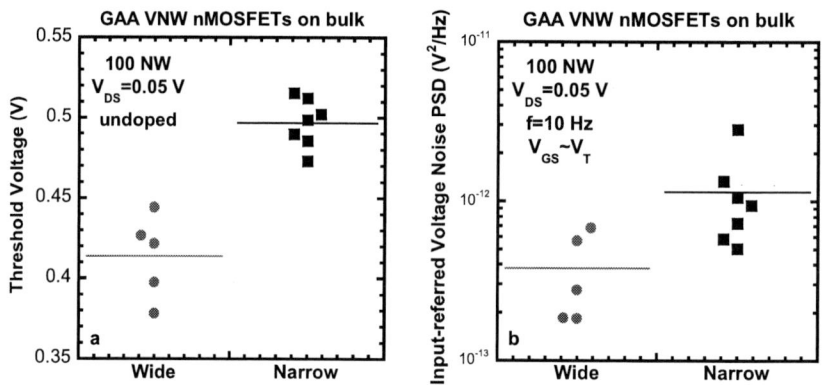

Figure 4. Threshold voltage (a) and input-referred noise PSD at 10 Hz (b) for a set of wide and narrow cap GAA VNW nMOSFETs without channel doping.

Impact of the contact asymmetry

VNW FETs on a silicon substrate, using the bulk as source contact exhibit pronounced asymmetry in the DC characteristics (5) that is related to the different series (or access) resistance at the drain when measuring in Forward (F) or Reverse (R) operation. In the latter case, the source and drain contacts have been switched. As shown recently, this asymmetry is also found in the LF noise spectra and PSD (6). An example of the S_{VG} versus V_{GS} at 10 Hz for a pFET on a bulk silicon wafer is shown in Fig. 5: a markedly lower 1/f noise PSD at 10 Hz is found in R compared with F operation. In both cases, the 1/f noise seems to be dominated by $\Delta\mu$ fluctuations, followed by an increase at higher I_D (more negative V_{GS}), due to series resistance. While the common $\Delta\mu$ origin of the 1/f noise suggests that conduction in the core of the pillar and away from the interface can be assumed, the difference in noise magnitude indicates a different carrier distribution (and/or number) leading to a different Hooge parameter α_H (30). The type of "injecting" source contact thus plays a crucial role in the transport across the silicon nanowire.

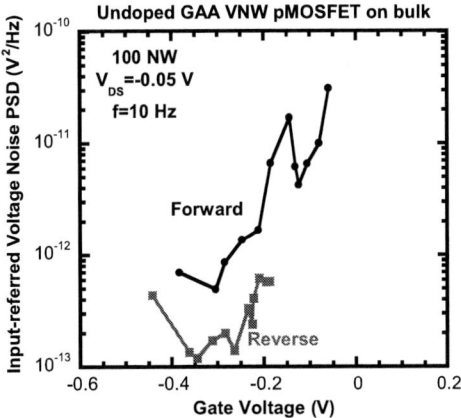

Figure 5. Input-referred voltage noise PSD at 10 Hz of an undoped GAA VNW pMOSFET on bulk in forward and reverse operation.

For a GAA VNW pFET on SOI on the other hand, little asymmetry can be found in the DC and noise characteristics (6). An example for an undoped nanowire pFET is given in Fig. 6, showing within the measurement accuracy similar S_{VG} values for F and R operation. Moreover, the 1/f noise mechanism in this case is rather Δn at small I_D in contrast to the $\Delta\mu$ fluctuations for the pFET on bulk silicon. A comparison of the input-referred voltage noise PSD in linear F operation and at 10 Hz is given in Fig. 7, showing a cross-over at intermediate I_D. In weak inversion, the 1/f noise PSD of the SOI pFET is lower, while it is opposite for the strong inversion regime. Moreover, it can be shown that no white noise is observed above 1 kHz for the VNW pFETs on SOI (35), indicating that at higher frequencies, these devices are favorable from a LF noise perspective compared with the ones on a bulk substrate. It also indicates that the substrate source contact originates the white noise and is not intrinsic to a GAA VNW FET.

Impact of the doping density
For the studied GAA VNW FETs the noise spectrum at low frequencies and in linear operation is dominated by flicker noise. In the case of the FETs with a bulk source contact, $\Delta\mu$ fluctuations originate the 1/f noise in subthreshold and low inversion operation. While for the nFETs on bulk, $\Delta\mu$ governs up to V_T, in the case of pFETs this is the case for most of the I_D-V_{GS} range studied here. Occasionally, Generation-Recombination (GR) noise giving rise to a Lorentzian spectral component has been observed (9). However, when single gate oxide traps, giving rise to Random Telegraph

Noise (RTN) can be dominant in a single nanowire, averaging over 100 parallel pillars yields predominantly 1/f noise.

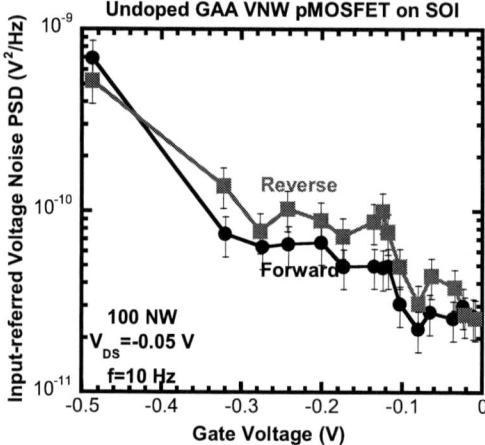

Figure 6. Input-referred voltage noise PSD at 10 Hz of an undoped GAA VNW pMOSFET on SOI in forward and reverse operation.

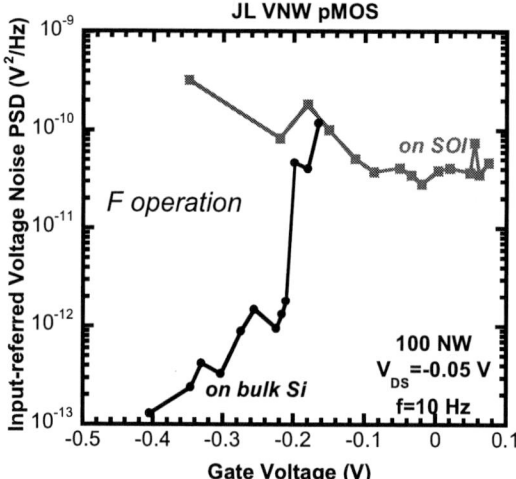

Figure 7. Input-referred voltage noise PSD at 10 Hz of an in-situ B-doped GAA VNW pMOSFET on bulk vs SOI in F operation.

For the VNW pFETs on SOI wafers, the situation regarding the nature of the 1/f noise is more complicated. As shown in Fig. 7 for undoped pillars, Δn fluctuations are present in weak inversion (low I_D), followed by $\Delta \mu$ correlated with number fluctuations and access resistance 1/f noise. However, the picture changes when increasing the in-situ B doping density denoted by [B] and corresponding with a junctionless architecture. This is illustrated by the input-referred voltage noise PSD versus V_{GS} represented in Fig. 8 for VNW pFETs on SOI with different [B]. Based on the gate voltage dependence in weak inversion, the 1/f noise is dominated by Δn for the lowest [B] of 2×10^{18} cm^{-3}, while for a [B] of 5 or 10×10^{18} cm^{-3}, $\Delta\mu$ fluctuations take over. This is further supported by the data in Fig. 9a to 9c, comparing the normalized drain current noise PSD (S_I/I_D^2) with $(g_m/I_D)^2$. The criterion for Δn dominated 1/f noise is that both characteristics run parallel, with the flat-band voltage noise PSD (S_{VFB0}) as proportionality constant (32), (33). For [B]=2×10^{18} cm^{-3} this is obeyed in weak inversion, resulting in an approximately constant S_{VFB0} up to -0.2 V in Fig. 8. No clear proportionality can be found for the higher pillar doping densities in Fig. 9b or 9c, suggesting a $\Delta\mu$ origin of the 1/f noise in weak inversion. Regarding the transport in the NW transistors, the 1/f noise results suggest that conduction moves from interface-dominated at low [B] to core-dominated at high(er) B doping concentration, for the subthreshold and weak inversion regimes.

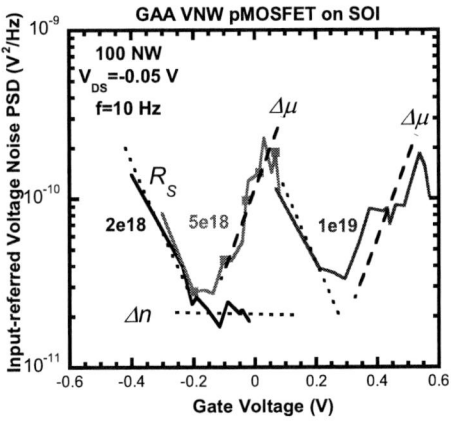

Figure 8. Input-referred voltage noise PSD at 10 Hz of in-situ B-doped JL GAA VNW pMOSFETs on SOI in F operation.

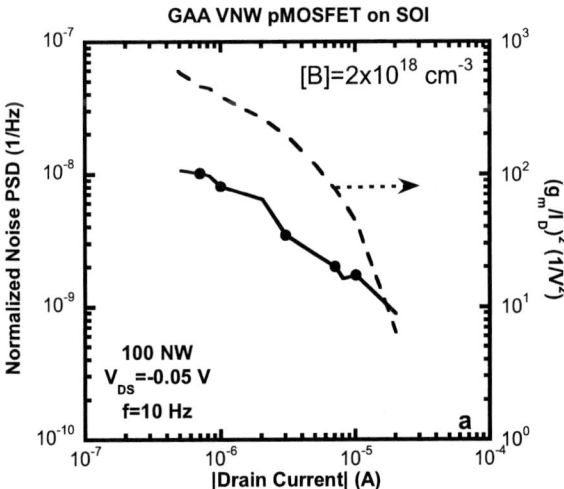

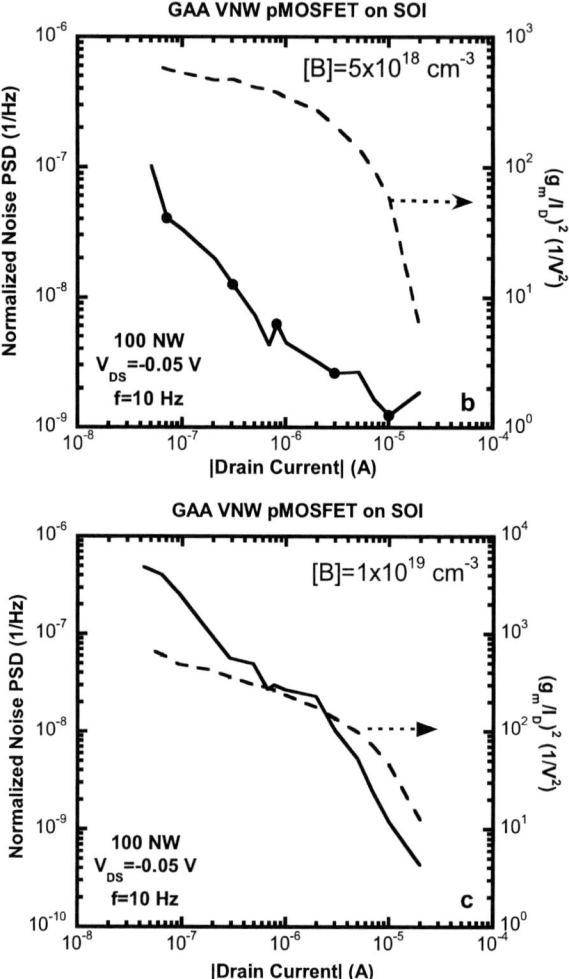

Figure 9. Normalized current noise PSD and $(g_m/I_D)^2$ versus absolute I_D in linear operation and f=10 Hz for a junctionless GAA VNW pFET with (a) [B]=2×10^{18} cm^{-3}; (b) [B]=5×10^{18} cm^{-3} and (c) [B]=1×10^{19} cm^{-3}.

Another interesting observation, better illustrated by Fig. 10, showing the average S_{VG} at 10 Hz and V_{DS}=-0.05 V is that the lowest values on average are found for the intermediate B concentration of 5×10^{18} cm^{-3}. This confirms earlier reports on the LF noise of junctionless, ion-implanted, horizontal GAA NW FETs (7), (9). In other words, the in-situ doping density of the silicon pillars is an important knob to optimize the LF noise performance of GAA VNW transistors.

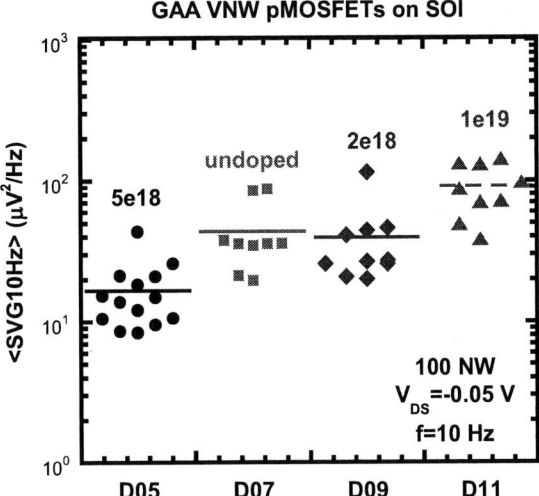

Figure 10. Average S_{VG} at 10 Hz and V_{DS}=-0.05 V for GAA VNW pFETs with different [B].

Summary

The impact of several process options on the LF noise of GAA VNW FETs has been described and discussed in detail. The biggest effcet has been noted for the type of source (backside bulk versus top contact). This not only affects the nature of the noise spectrum, with a pronounced white noise (>1 kHz) following the dominant flicker noise below 1 kHz but also introduces a significant asymmetry in the device characteristics. The origin of the 1/f noise in this case are mobility fluctuations, dominating in subthreshold and weak inversion for both n- and p-type FETs. In the case of a 3D implementation of VNW FETs, a local source contact will be employed, suppressing the asymmetry and the white noise. At the same time, the dominant 1/f noise originates from number fluctuations for the subthreshold and weak inversion regimes, in the case of undoped or lowly doped pillars. For increasing in-situ B doping density, the flicker noise in the junctionless VNW pFETs becomes progressively more governed by mobility fluctuations. This can be explained by a shift of the conduction path from near the Si/SiO_2 interface to the interior of the nanowire. At the same time, optimal LF noise performance can be obtained for an intermediate [B] in the range of 5×10^{18} cm^{-3}.

Acknowledgments

This work has been performed in the Partner Program on GAA Logic Devices.

References

1. A. Veloso, M. J. Cho, E. Simoen, G. Hellings, P. Matagne, N. Collaert and A. Thean, *ECS Trans.*, **72** (2), 85 (2016).
2. A. Veloso, P. Matagne, E. Simoen, B. Kaczer, G. Eneman, H. Mertens, D. Yakimets and B. Parvais, *J. Phys: Condens. Matter*, **30**, 384002 (2018).
3. A. Veloso, G. Eneman, A. De Keersgieter, D. Jang, H. Mertens, P. Matagne, E. Dentoni Litta, J. Ryckaert and N. Horiguchi, in *Proc. of the 5th IEEE Electron Devices Technol. & Manufact. Conf. (EDTM)*, IEEE Xplore, p. 1 (2021).
4. A. Veloso, G. Eneman, T. Huynh-Bao, A. Chasin, E. Simoen, E. Vecchio, K. Devriendt, S. Brus, E. Rosseel, A. Hikavyy, R. Loo, V. Paraschiv, B. T. Chan, D. Radisic, W. Li, J. J. Versluijs, L. Teugels, F. Sebaai, P. Favia, H. Bender, E. Vancoille, J. Scheerder, C. Fleischmann, N. Horiguchi and P. Matagne, in the *Proc. of IEDM19*, San Francisco (CA), IEEE Xplore, p. 230 (2019).
5. M. Liu, F. Lentz, S. Trellenkamp, J.-M. Hartmann, J. Knoch, D. Grützmacher, D. Buca and Q.-T. Zhao, *IEEE Trans. Electron Devices*, **67**, 2988 (2020).
6. E. Simoen, A. Veloso and P. Matagne, to be presented at *EUROSOI-ULIS 2021*, Caen (France), 1-3 Sept. 2021.
7. A. Veloso, G. Hellings, M. J. Cho, E. Simoen, K. Devriendt, V. Paraschiv, E. Vecchio, Z. Tao, J. J. Versluijs, L. Souriau, H. Dekkers, S. Brus, J. Geypen, P. Lagrain, H. Bender, G. Eneman, P. Matagne, A. De Keersgieter, W. Fang, N. Collaert and A. Thean, in *Tech. Dig. of the Symposium on VLSI Technol.*, IEEE Xplore, p. T138 (2015).
8. E. Simoen, A. Veloso, P. Matagne, N. Collaert and C. Claeys, *IEEE Trans. Electron Devices*, **65**, 1487 (2018).
9. E. Simoen, A. Vinicius de Oliveira, P. Ghedini Der Agopian, R. Ritzenthaler, H. Mertens, N. Horiguchi, J. Antonio Martino, C. Claeys and A. Veloso, *Solid-State Electron*, **184**, 108087 (2021).
10. Y. F. Lim, Y. Z. Xiong, N. Singh, R. Yang, Y. Jiang, D. S. H. Chan, W. Y. Loh, L. K. Bera, G. Q. Lo, N. Balasubramanian and D.-L. Kwong, *IEEE Electron. Device Lett.*, **27**, 765 (2006).
11. C. Wei, Y.-Z. Xiong, X. Zhou, N. Singh, S. C. Rustagi, G. Q. Lo and D.-L. Kwong, *IEEE Electron Device Lett.*, **30**, 668 (2009).
12. R.-H. Baek, C.-K. Baek, H.-S. Choi, J.-S. Lee, Y. Y. Yeoh, K. H. Yeo, D.-W. Kim, K. Kim, D. M. Kim and Y.-H. Jeong, *IEEE Trans. Nanotechnol.*, **10**, 417 (2011).
13. W. Fang, A. Veloso, E. Simoen, M.-J. Cho, N. Collaert, A. Thean, J. Luo, C. Zhao, T. Ye and C. Claeys, *IEEE Electron Device Lett.*, **37**, 363 (2016).
14. P. Singh, N. Singh, J. Miao, W.-T. Park and D. L. Kwong, *IEEE Electron Device Lett.*, **32**, 1752 (2011).
15. C.-H. Park, M.-D. Ko, K.-H. Kim, S.-H. Lee, J.-S. Yoon, J.-S. Lee and Y.-H. Jeong, *IEEE Electron Device Lett.*, **33**, 1538 (2012).
16. J. Zhuge, R. Wang, R. Huang, Y. Tian, L. Zhang, D.-W. Kim, D. Park and Y. Wang, *IEEE Electron Device Lett.*, **30**, 57 (2009).
17. N. Clément, X. L. Han and G. Larrieu, *Appl. Phys. Lett.*, **103**, p. 263504 (2013).
18. T. Imamoto, Y. Ma, M. Muraguchi and T. Endoh, *Jpn. J. Appl. Phys.*, **54**, 04DC11 (2015).
19. C. Mukherjee, C. Maneux, J. Pezard and G. Larrieu, in *Proc. ESSDERC 2017*, IEEE Xplore, p. 34 (2017).

20. U.-S. Jeong, C.-K. Kim, H. Bae, D.-I. Moon, T. Bang, J.-M. Choi, J. Hur and Y.-K. Choi, *IEEE Trans. Electron Devices*, **63**, 2210 (2016).
21. T. Bang, B.-H. Lee, C.-K. Kim, D.-C. Ahn, S.-B. Jeon, M.-H. Kang, J.-S. Oh and Y.-K. Choi, *IEEE Electron Device Lett.*, **38**, 40 (2017).
22. S.-W. Lee, T. Bang, C.-K. Kim, K.-M. Hwang, B. C. Jang, D.-I. Moon, H. Bae, M. Seo, S.-Y. Kim, D.-H. Kim, S.-Y. Choi and Y.-K. Choi, *IEEE Electron. Device Lett.*, **38**, 1008 (2017).
23. W. Feng, R. Hettiarachchi, S. Sato, K. Kakushima, M. Niwa, H. Iwai, K. Yamada and K. Ohmori, *Jpn. J. Appl. Phys.*, **51**, 04DC06 (2012).
24. K. Ohmori, W. Feng, R. Hettiarachchi, Y. Lee, S. Sato, K. Kakushima, M. Sato, K. Fukuda, M. Niwa, K. Yamabe, K. Shiraishi, H. Iwai and K. Yamada, *ECS Trans.*, **45** (3), 437 (2012).
25. S.-H. Lee, C.-K. Baek, S. Park, D.-W. Kim, D. K. Sohn, J.-S. Lee, D. M. Kim and Y.-H. Jeong, *IEEE Electron Device Lett.*, **33**, 1348 (2012).
26. C. Liu, R. Wang, J. Zou, R. Huang, C. Fan, L. Zhang, J. Fan, Y. Ai and Y. Wang, in *IEDM11 Tech. Dig.*, IEEE Xplore, p. 521 (2011).
27. W. Feng, R. Hettiarachchi, Y. Lee, S. Sato, K. Kakushima, M. Sato, K. Fukuda, M. Niwa, K. Yamabe, K. Shiraishi, H. Iwai and K. Ohmori, in *IEDM11 Tech. Dig.*, IEEE Xplore, p. 630 (2011).
28. J. Zhuge, L. Zhang, R. Wang, R. Huang, D.-W. Kim, D. Park and Y. Wang, *Appl. Phys. Lett.*, **94**, 083503 (2009).
29. E. Simoen, A. Chasin, P. Matagne, E. Rosseel, A. Hikavyy, R. Loo, P. Favia, H. Bender, E. Vancoille and A. Veloso, *ECS Trans.*, **97** (5), 59 (2020).
30. F. N. Hooge, T. G. M. Kleinpenning and L. K. J. Vandamme, *Rep. Prog. Phys.*, **44**, 479 (1981).
31. E. Simoen, and C. Claeys, *Solid-State Electronics*, **43**, 865 (1999).
32. G. Ghibaudo, O. Roux, Ch. Nguyen-Duc, F. Balestra and J. Brini, *Phys. Status Solidi (a)*, **124**, 571 (1991).
33. G. Ghibaudo and T. Boutchacha, *Microelectron. Reliab.*, **42**, 573 (2002).
34. E. Simoen, H.-C. Lin, A. Alian, G. Brammertz, C. Merckling, J. Mitard and C. Claeys, *IEEE Trans. Device and Mater. Reliability*, **13**, 444 (2013).
35. E. Simoen, A. Veloso, B. O'Sullivan, K. Takakura and C. Claeys, in *Proc. of the 2021 China Semiconductor Technology International Conference (CSTIC)*, IEEE Xplore, p. 1 (2021).

14

Chapter 2

Advanced Device Technology 2

16

HfZrO-based Ferroelectric devices for lower power AI and memory applications

Shinichi Takagi, Kasidit Toprasertpong, Kento Tahara, Eishin Nako, Ryosho Nakane, Zeyu Wang, Xuan Luo, Tsung-En Lee and Mitsuru Takenaka

Department of Electrical Engineering and Information Systems, the University of Tokyo, Tokyo, Japan

Si-friendly HfO_2-based ferroelectric devices have been strongly recognized as a novel technology booster for future integrated memory and logic systems. In this paper, we address our recent activities on $TiN/Hf_{0.5}Zr_{0.5}O_2(HZO)/TiN$ MFM capacitors, HZO/Si FeFETs for memory applications and a newly-proposed reservoir computing using HZO/Si FeFETs and MFM capacitors for AI applications. We have demonstrated that MFM capacitors with HZO less than 5 nm can realize low crystallization temperature, excellent ferroelectricity, low operating voltage and high read/write endurance by performing sufficient wake-up operations to the thin HZO films. For the FeFET memory, we have found the importance of interfacial layers (ILs) between HZO and Si on the memory window. It has been revealed that the IL thickness is sensitive to the process temperature and that electron trapping around HZO/ILs has significant impacts on the memory operation. Finally, we have proposed and experimentally demonstrated reservoir computing using FeFETs for neuromorphic applications.

Introduction

Electron devices using ferroelectric materials have recently stirred a strong interest as a technology booster for future integrated systems, since the discovery of HfO_2-based ferroelectrics [1], which are CMOS-friendly and scalable. At present, a variety of devices using HfO_2-based ferroelectrics such as FeRAM, FeFETs and FTJ have been extensively studied for low power memories and computing-in-memory applications [2]. This enthusiasm has been enhanced by the increasing demands and strong concerns of AI calculations, which needs coexistence of both logic and memory functions in an identical device. Thus, we are currently investigating the possibilities of FeRAM and FeFETs using $Hf_{0.5}Zr_{0.5}O_2$ (HZO) films from the viewpoints of ultralow power memory and AI applications. In this paper, we report the recent research activities on impact of HZO materials and fabrication processes on performance of ferroelectric devices using HZO as well as a new approach to AI calculations by reservoir computing using FeFETs and FeRAM.

Characterization of MFM structures and FeRAM applications

For FeRAM, we have examined the possibility of low voltage operation by thinning HZO films under low thermal budget needed for formation in back-end-of-line (BEOL) for future non-volatile memory applications [3]. This is because the low voltage operation allows FeRAM to be embedded with logic LSIs under advanced technology nodes and is expected to improve the endurance characteristics, which are regarded as one of the most difficult challenges of HZO-based memories. Thus, we systematically investigated the ferroelectric characteristics of $Hf_{0.5}Zr_{0.5}O_2$ films as a parameter of annealing temperature and the HZO thickness from 9.5 nm down to 2.8 nm. Fig. 1 (a) shows the P-E characteristics of TiN/4.0 nm-thick HZO/TiN capacitors annealed at 500 °C, taken under the pristine condition, after wake-up of 10^6 cycles at 2.5 MV/cm and after wake-up of 10^6 cycles at 4 MV/cm. It is observed that the ferroelectric properties of thin pristine films are quite poor. On the other hand, the ferroelectricity is activated after strong-field (4 MV/cm) cycling and the thin HZO films exhibit clear hysteresis loops even in operating field much lower than the cycling field, (<< 4 MV/cm), while the low-field cycling is insufficient to obtain hysteresis loops with wide memory window. This fact indicates that MFM capacitors with thin HZO films can be still used effectively by utilizing strong-field cycling.

We have established the thickness-temperature mapping of 2Pr values after strong-field cycling to show a tradeoff relationship between the thickness scaling and crystallization temperature, which has to be taken into account in the implementation as BEOL FeRAM. Fig. 1(b) shows the ferroelectric characteristics of TiN/HZO/TiN capacitors as functions of annealing temperature and the HZO thicknesses. Here, ● and x in the map mean ferroelectricity with 2Pr > 10 $\mu C/cm^2$ and < 10 $\mu C/cm^2$, respectively. Note that the 2Pr values shown here are the ones after cycling. It is observed that annealing temperature sufficient to obtain 2Pr values becomes higher with reducing the HZO thickness, indicating the existence of a clear tradeoff between ferroelectric-phase crystallization temperature and the HZO thickness less than 6 nm. Even under this trade-off, however, we have demonstrated that MFM capacitors with HZO less than 5 nm can realize crystallization temperature of 500 °C and lower, and the excellent ferroelectricity of 2Pr > 25 $\mu C/cm^2$.

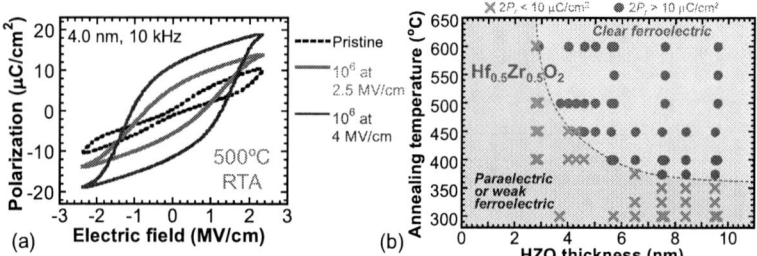

Fig. 1. (a) P-E characteristics of 4.0 nm/500°C HZO at 2.5 MV/cm (1.0 V) for the pristine state, after wake-up at 2.5 MV/cm for 10^6 cycles, and after wake-up at 4 MV/cm for 10^6 cycles. (b) Thickness-annealing temperature mapping for confirmed ferroelectricity with 2Pr > 10 $\mu C/cm^2$ in TiN/HZO/TiN capacitors.

Such thinner HZO films are favorable to achieve sufficient field against the coercive field, Ec, for polarization reversal at low operating voltage. While the thick HZO cannot perform well at applied voltage below 1.2 V, 4.0-nm and 4.6-nm HZO films exhibit excellent ferroelectric hysteresis loops down to 0.7 V and 1.0 V. We have confirmed the comparatively better data retention in the thin HZO under low-voltage operation. At 1.2 V and 1.0 V read/write operations, much superior same-state (SS) data retention is found in the 4.0-nm and 4.6-nm HZO films than in the 6.5 nm ones, attributed to the higher read/write field.

On the other hand, thinner HZO films are expected to improve the reliability of MFM capacitors. We have observed that the breakdown field, E_{bd}, is 3.5 MV/cm in 9.5-nm HZO and reaches 6 MV/cm in 4.0-nm HZO, indicating the higher E_{bd} in thinner HZO. This might be attributable to the decrease in the energy of electrons flowing through and the resulting less damages to the HZO layer [2], as similar to the thin SiO_2/Si MOS system with improved breakdown characteristics [4, 5]. This E_{bd} increase with thickness scaling can mitigate the problem in breakdown-limited endurance, originating in the low E_{bd}/E_C ratio in HZO films. We have found that a 4.0-nm HZO capacitor with the area of 3600 μm^2 has endurance of as high as 10^{10} under the high cycling field of 4 MV/cm at 100 kHz with keeping $2Pr > 20$ $\mu C/cm^2$. When reducing the cycling field to 3 MV/cm (1.2 V operation) after wake-up, the 4.0-nm HZO capacitor does not cause breakdown within the experimental stress time of 10^{10} cycles. This result suggests endurance with cycles higher than 10^{14} under a decreased applied field of 3 MV/cm and increased cycling frequency of 10 MHz. In addition, further >1000x improvement could be expected by the area scaling [6]. We have found that the fatigue-limited endurance of the 4.0-nm HZO MFM capacitor with wake-up at 4 MV/cm for 10^6 cycles can also be extrapolated to over 10^{14} even at 3 MV/cm (1.2 V operation). These results, in turn, indicate that the high-field (4 MV/cm) wake-up cycling for 10^6 cycles can help improvement in the ferroelectricity of ultra-thin HZO without reliability penalty. As a result, we have demonstrated that MFM capacitors with HZO less than 5 nm can realize crystallization temperature lower than 500 °C, excellent ferroelectricity ($2Pr > 25$ $\mu C/cm^2$), low operating voltage (0.7-1.2 V), and high read/write endurance projected to 10^{14} cycles by performing sufficient amounts of wake-up operations to the thin HZO films.

Understanding of electrical characteristics of FeFET

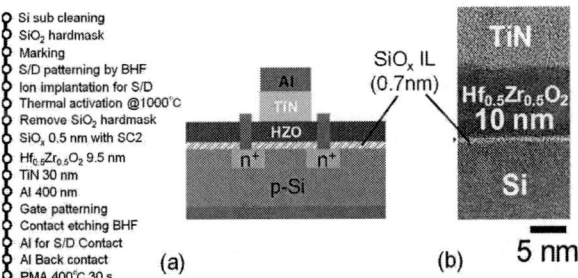

Fig. 2. (a) process flow and device structure of a HZO FeFET and (b) cross sectional TEM picture of a fabricated HZO FeFET.

HZO-based FeFETs are currently expected for a variety of applications [2] including negative capacitance FETs, high drive-current logic devices, embedded memories, cross-bar Commutation-in-Memory (CiM) [7], multiply-accumulate (MAC) operation CiM [8,

9] and so on. On the other hand, the operation of FeFETs has not been fully understood yet. Also, one of the critical issues of FeFETs is degradation of FeFETs with cycling operations and resulting low endurance, which is attributed mainly to ferroelectric/ semiconductor interfaces and interfacial layers (ILs). However, impacts of ferroelectric/ semiconductor interfaces on FeFET characteristics and reliability have not been clarified yet. Thus, we have re-examined the impact of the fabrication process and the operation of FeFETs [10-13].

The typical fabrication process flow is shown in Fig. 2(a). A p-Si substrate with N_{sub} of 4×10^{15} cm^{-3} was cleaned and chemically oxidized to form a thin SiO_2 IL. 10-nm-thick ferroelectric $Hf_{0.5}Zr_{0.5}O_2$ was then stacked by ALD before capped by TiN electrode. After the FET fabrication process, the device was annealed at 400°C for 30 seconds in N_2 atmosphere to form the ferroelectric phase in $Hf_{0.5}Zr_{0.5}O_2$. A TEM image in Fig. 2(b) shows 0.7-nm-thick SiO_2 IL at the Si interface. Fig. 3(a) shows the P-E characteristics of a TiN/HZO (10 nm)/TiN MFM capacitor, fabricated under the same condition as in FeFETs. Good ferroelectric characteristics with $2Pr > 40$ μC/cm^2 are confirmed in the present HZO films. Fig. 3(b) shows the I_d-V_g and I_g-V_g characteristics of a fabricated HZO FeFET. A clear memory window of around 2 V is confirmed.

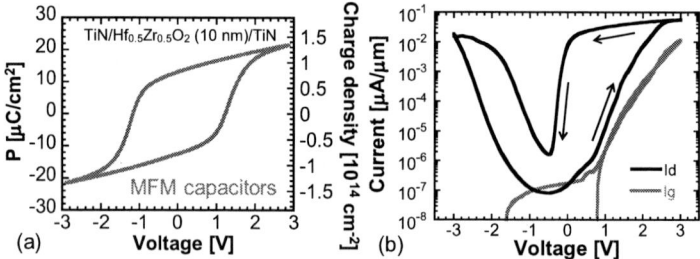

Fig. 3. (a) P-E characteristics of a TiN/HZO (10 nm)/TiN MFM capacitor and (b) I_d-V_g and I_g-V_g characteristics of a fabricated HZO FeFET.

On the other hand, we have found that the memory window and the sub-threshold swing (S.S.) of FeFETs are dependent on PMA temperature [13]. Fig. 4 summarizes the memory windows, estimated from the threshold voltages at forward and backward scan defined by the constant current of $(10^{-7} \times W/L)$ A, as a function of PMA temperature. The amount of memory window is maximized around 500 °C. The decrease in the memory window with further increasing temperature is attributed to the higher drop across IL caused by increasing the IL thickness and higher electron trapping due to the degradation of the ferroelectric/semiconductor interfaces. Actually, the thicker IL with PMA temperature higher than 500 °C is confirmed by TEM observation. The average subthreshold swing S.S. is also shown in Fig. 4. The increasing annealing temperature results in the degradation of S.S. due to the increase in D_{it}. PMA temperature around 600-700 °C has been well reported for HZO MFM capacitors in terms of sufficient crystallization of HZO films. Judging from the present results, however, we can conclude that much lower PMA temperature around 500 °C should be chosen for FeFET fabrications in order to achieve both good crystallinity of HZO films and HZO/Si MOS interface properties.

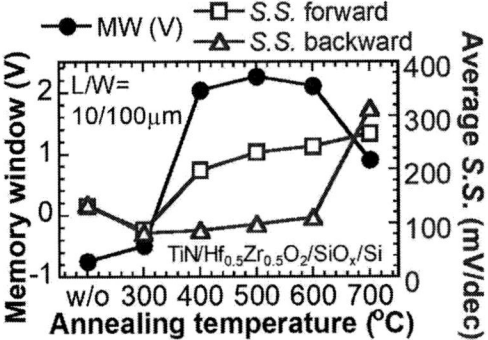

Fig. 4. Memory window and average S.S. of FeFETs annealed at different temperatures.

Furthermore, we have clarified that a large amount of electron trapping near the HZO/interfacial layer interfaces can help polarization of HZO while hole trapping is less significant [10, 11]. We have discriminated areal concentrations of free electrons and holes in the channels from the areal trapped charges by employing quasi-static split C-V and Hall measurements to Si n- and p-channel FeFETs. The schematic picture of carrier distributions in n-channel FeFETs with high positive gate voltage is shown in Fig. 5. A large amount of electrons trapped at HZO/IL interfaces leads to the increase in electric field across the HZO films and a resulting high amount of positive polarization, while the areal density of free electrons in the inversion-layer of n-channel FETs is much reduced. On the other hand, such a large amount of hole traps does not exist and, thus, the electric field across the HZO films in p-channel FeFETs with negative gate voltage becomes much smaller than that in n-channel ones. As a result, the amount of negative polarization in p-channel FeFETs with negative gate voltage is much lower than that of positive polarization in n-channel FeFETs with positive gate voltage, inducing asymmetric P-V loops in the FeFET memory operation. The electron trapping properties in the HZO/Si MFIS structures can significantly affect the memory window, the read-out operation and the FeFET reliability.

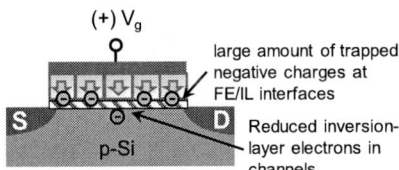

Fig. 5. Charge distribution during ON operation of an n-channel FeFET.

Reservoir computing using ferroelectric-based devices

As introduced in the previous section, FeFETs are strongly expected for CiM circuits for AI applications, because FeFETs do hold both logic and memory functions in an identical device. On the other hand, the complicated dynamics of the FeFET operation including polarization switching and carrier trapping can provide a possibility to utilize this device as AI calculation hardware in a different way from the CiM circuits such as a cross-bar memory array [7] and a MAC function array [8, 9]. Actually, we have proposed a new AI calculation scheme by reservoir computing using FeFETs for neuromorphic applications [14, 15].

Here, reservoir computing [16, 17] is a computational framework that can extract history-dependent information in time-series data and its remarkable feature is extremely light training computational load. In reservoir computing, time-series inputs are nonlinearly transformed to high-dimensional output signals by a reservoir that is modeled as a recurrent neural network with sparse and fixed weight connections between internal units. Owing to such a role of the reservoir, adaptive weights of the reservoir-to-output connections are adjusted by linear regression learning algorithms, which enables high-speed learning [18]. Thus, reservoir computing has recently been attracting a strong attention as a method to realize real-time processing of learning functions (real-time learning) in AI calculations at the edge. Furthermore, when a reservoir is represented by hardware that has input-history-dependent and nonlinear dynamics (physical reservoir), as shown in Fig. 6, we can expect further advantages in terms of power consumption and computational load, leading to significant improvement in energy efficiency of AI calculations.

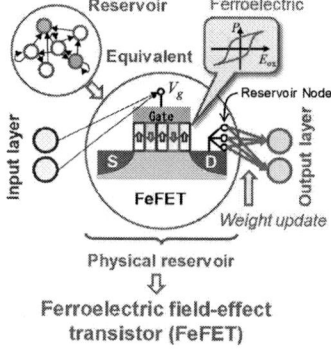

Fig. 6. Proposed physical reservoir computing by using FeFETs.

We have recently proposed and demonstrated a method using HfO_2-based FeFETs and MFM capacitors with temporary memory and nonlinearity as this physical reservoir (Fig. 6) [14, 15, 19-22]. The advantages of HfO_2-based ferroelectric devices as a physical reservoir are as follows. (1) The memory effect, polarization, domain interaction, and complex time response of ferroelectric thin films, and their nonlinear output readout as transient currents in the device, are expected to provide various internal states, rich response characteristics, and rich dynamics required for physical reservoirs, as schematically shown in Fig. 7. HfO_2-based ferroelectric devices, which are highly compatible with the Si CMOS process and have been used in advanced logic LSIs, can be easily integrated with existing CMOS-based circuit systems, which are the standard platform for AI calculation. It is also expected to improve the system performance by seamlessly coupling with the input/output processing of Si CMOS circuits.

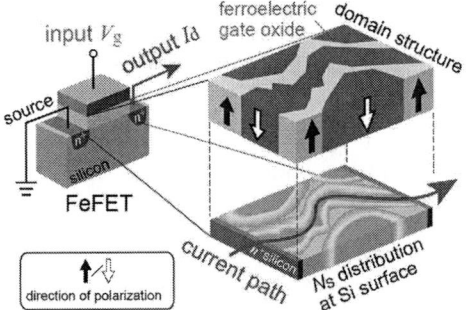

Fig. 7. Schematic view of time-dependent non-linear response of channel currents in FeFETs for time series gate voltage input.

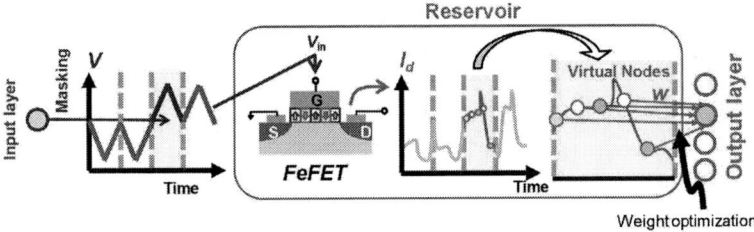

Fig. 8. Proposed reservoir computing scheme utilizing time responses of drain current of FeFETs as virtual nodes of neural networks.

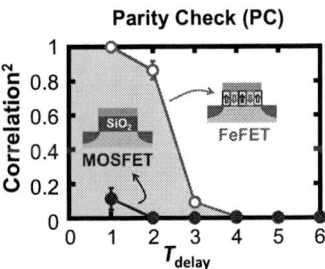

Fig. 9. Experimental correlation between the output and test data in parity check task.

The schematic diagram of our proposed FeFET-based reservoir system is shown in Fig. 8. A time-series signal defined by a fixed time step is input to the gate, and the drain current at each time sampled in the step is used as the virtual output node state, and the result of weighting is used as the system output. By learning the value of these weights according to a specific task, inference for unknown inputs becomes possible. In the present work, triangular waveforms, whose positive/negative peaks correspond to 1/0 digital inputs, were used as input V_g signals.

In this study, we have performed the Parity Check (PC) task to examine both the memory and the non-linearity of the input data at a previous time step (T_{delay}) [23], which are defined by

$$y_n^{(PC)} = S_{n-T_{delay}} + S_{n-T_{delay}+1} \cdots + S_n \; (mod \; 2)$$ [1]

The performance of the PC task was evaluated by correlations between input (test) values and inferred values, determined by

$$Cor\left(T_{delay}\right)^2 = \frac{Cov\left[y_{test}\left(t,T_{delay}\right),y_{out}(t)\right]^2}{Var\left[y_{test}\left(t,T_{delay}\right)\right]\cdot Var\left[y_{out}(t)\right]}$$ [2]

as a function of T_{delay}. Fig. 9 shows the experimental results of the inference performance of the PC task between $Hf_{0.5}Zr_{0.5}O_2$/Si FeFETs and conventional SiO_2/Si MOSFETs. It is found that FeFETs exhibit the performance up to T_{delay} of 3, which is different from the results of SiO_2/Si MOSFETs. This fact indicates that polarization switch in FeFETs contributes to reservoir computing. We have also confirmed that the similar reservoir computing can be performed in TiN/ $Hf_{0.5}Zr_{0.5}O_2$/TiN MFM capacitors by utilizing the polarization current associated with polarization switching [19].

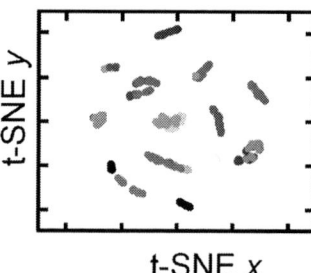

Fig. 10. t-SNE analyses of experimental results for 4-bit input data.

The T-distributed Stochastic Neighbor Embedding (t-SNE) method [24] was used to evaluate the data identification performance of the present reservoir computing using FeFETs for 4-bit digital inputs. Fig. 10 shows the results of evaluating the data discrimination performance for a 4-bit digital input using the t-SNE method. It can be seen that a single FeFET can discriminate between 12-13 different input patterns.

From the above results, physical reservoirs with ferroelectrics such as FeFETs are promising as a new reservoir computing adapted to the Si platform, and are expected to be applied to real time AI processing of time series data at the edge.

Conclusions

In this paper, we have addressed a variety of aspects of HZO-based ferroelectric devices from memory applications to AI calculations. Also, we have emphasized the importance of physical understanding of device/material structures and their electrical characteristics, which will contribute to further enhancement of the performance and the reliability of these HZO-based ferroelectric FETs and MFM capacitors.

Acknowledgments

This work was supported by JST CREST (JPMJCR20C3), JSPS KAKENHI (21H01359), and Nanotechnology Platform project by MEXT (JPMXP09A20UT0056), Japan.

References

1. T. S. Böscke, J. Müller, D. Bräuhaus, U. Schröder and U. Böttger, *Appl. Phys. Lett.* **99**, 102903 (2011)
2. A. I. Khan, A. Keshavarzi and S. Datta, *Nature Electronics* **3**, 588 (2020).
3. K. Tahara, K. Toprasertpong, Y. Hikosaka, K. Nakamura, H. Saito, M. Takenaka and S. Takagi, *IEEE Symp. on VLSI Technol.*, T7-3 (2021).
4. K. F. Schuegraf and C. Hu., *IEEE Trans. Electron Devices*, **41**, 761 (1994).
5. J. H. Stathis and D. J. DiMaria, *Tech. Dig. IEEE Int. Electron Device Meeting*, 167 (1998).
6. L. Grenouillet, T. Francois, J. Coignus, S. Kerdilès, N. Vaxelaire, C. Carabasse, F. Mehmood, S. Chevalliez, C. Pellissier, F. Triozon, F. Mazen, G. Rodriguez, T. Magis, V. Havel, S. Slesazeck, F. Gaillard, U. Schroeder, T. Mikolajick, E. Nowak, *IEEE Symp. on VLSI Technol.*, TF2.4 (2020).
7. M. Jerry, P.-Y. Chen, J. Zhang, P. Sharma, K. Ni, S. Yu and S. Datta, *Tech. Dig. IEEE Int. Electron Device Meeting*, 139 (2017).
8. K. Kamimura, S. Nohmi, K. Suzuki and K. Takeuchi, *Tech. Dig. IEEE European Solid-State Device Research Conference (ESSDERC)*, 178, (2019).
9. C. Matsui, K. Toprasertpong, S. Takagi, and K. Takeuchi, *IEEE Symp. on VLSI Technol.*, JFS2-8 (2021).
10. K. Toprasertpong, M. Takenaka, and S. Takagi, *Tech. Dig. IEEE Int. Electron Device Meeting*, 570 (2019).
11. K. Toprasertpong, Z. -Y. Lin, T. -E. Lee, M. Takenaka, and S. Takagi, *IEEE Symp. on VLSI Technol.*, TF1.5 (2020).
12. K. Toprasertpong, K. Tahara, M. Takenaka, and S. Takagi, *Appl. Phys. Lett.*, **116**, 242903 (2020).
13. K. Toprasertpong, K. Tahara, T. Fukui, Z. Lin, K. Watanabe, M. Takenaka and S. Takagi, *IEEE Electron Device Lett.*, **41**, 1588 (2020).
14. E. Nako, K. Toprasertpong, R. Nakane, Z. Wang, Y. Miyatake, M. Takenaka, and S. Takagi, *IEEE Symp. on VLSI Technol.*, TN1.6 (2020).
15. K. Toprasertpong, E. Nako, R. Nakane, Z. Wang, Y. Miyatake, M. Takenaka and S. Takagi, *International Symposium on Nonlinear Theory and Its Applications (NOLTA)*, 458 (2020).
16. W. Maass, T. Natschläger, and H. Markram, *Neural Comp.* **14**, 2531 (2002).

17. H. Jaeger, *GMD Technical Report* 148 (2001).
18. G. Tanaka, T. Yamane, J. B. Hérouxc, R. Nakane, N. Kanazawa, S. Takeda, H. Numata, D. Nakano, A. Hirose, *Neural Networks* **115**, 100 (2019).
19. E. Nako, K. Toprasertpong, R. Nakane, Z. Wang, M. Takenaka and S. Takagi, *International Conference on Solid State Devices and Materials (SSDM)*, 99-100 (2020).
20. Z. Wang, K. Toprasertpong, Z. Lin, E. Nako, R. Nakane, M. Takenaka and S. Takagi, *Silicon Nanoelectronics Workshop (SNW)*, S3-3 (2021).
21. E. Nako, K. Toprasertpong, R. Nakane, Z. Wang, M. Takenaka and S. Takagi, presented in *International Conference on Solid State Devices and Materials (SSDM)* (2021).
22. Z. Wang, E. Nako, K. Toprasertpong, R. Nakane, M. Takenaka and S. Takagi, presented in *International Conference on Solid State Devices and Materials (SSDM)* (2021).
23. T. Furuta, K. Fujii, K. Nakajima, S. Tsunegi, H. Kubota, Y. Suzuki, and S. Miwa, *Phys. Rev. Applied*, **10**, 034063 (2018).
24. L. van der Maaten and G. Hinton, *J. Mach. Learn. Research*, **9**, 2579 (2008).

Vertically Stacked Junction Devices Fabricated Using Single-Crystal Graphene on SiC Substrate

M. Nagase

Institute of Post-LED Photonics, Tokushima University, Tokushima 770-8506, JAPAN

Two types of vertically stacked graphene junction diodes were fabricated in this study. Samples of single-crystal graphene measuring 100 mm^2 were epitaxially grown on SiC substrate using the thermal decomposition method and were bonded using the direct bonding technique. The direct-bonded stacked junction diode exhibited nonlinear current-voltage characteristics and acted as a far-infrared emitter. Fowler-Nordheim tunneling phenomena with a strong nonlinear behavior was observed in the tunneling diode with a thin insulative layer (air gap or structured water). By using simple device-assembly processes, vertically stacked graphene diodes with new functions were successfully fabricated.

Introduction

Graphene is predicted to become a fundamental component in future electronic devices because of its exceptional electrical properties, such as high mobility and unique band structure (1). Recently, vertically stacked graphene junctions have attracted significant attention because they exhibit properties such as superconductivity (2). Generally, stacked graphene devices are fabricated using graphene flake or poly-crystal graphene. Because a complex assembly process is required, the reproducibility of these devices is low. Therefore, we propose a fabrication process for the stacked graphene devices based on the direct bonding technique. The samples of single-crystal graphene grown on SiC substrate were bonded in a face-to-face configuration. The contact pins were in direct contact with the graphene layer to form electrical connections. A far-infrared emitter diode and tunneling diode were fabricated using this simple method.

Results and Discussion

Two types of stacked graphene junction diodes are discussed in this paper − the directly bonded graphene-graphene junction diode, shown in Fig. 1(a), and tunneling diodes with a tunneling barrier, depicted in Fig. 1(b).

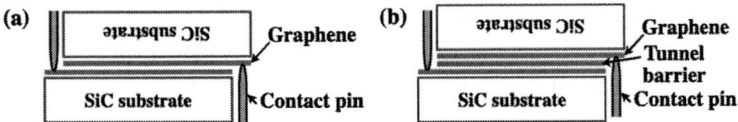

Figure 1. Schematic of vertically stacked junction device.
(a) graphene-graphene junction diode. (b) tunneling diode.

Graphene on SiC Substrate

The single-crystal graphene was epitaxially grown on a semi-insulating SiC (0001) substrate using the thermal decomposition method (3). The diced sample measuring 100 mm^2 was annealed in a rapid thermal annealer (SR-1800: Thermo-Riko). The crystals were grown at 1600 °C for 5 min in Ar (100 Torr). Surface roughness (a crucial factor to realize direct bonding) below 1 nm was ensured using the surface structure control technique.

Fabrication of Device

Two graphene samples were bonded with each other in a face-to-face manner and fixed with an acrylic mold. The gold-coated contact pins were in direct contact with the graphene surface. Because the contact resistance of epitaxial graphene on SiC was significantly smaller (4) than that of other graphene types (such as chemical vapor deposited graphene), ohmic contact was easily established by assembling the contact pins without using lithographic techniques. An air gap was used as a tunneling barrier, necessary for a tunneling diode, in the softly bonded samples. Here, the ohmic contact between two the graphene layers was established by increasing the bonding load. A structure water layer (5) formed using deionized water (DI) treatment was also used as a tunneling barrier. As the thickness of the structured water layer was less than 1 nm, it was suitable for the tunneling diode. The fabrication procedures for direct contact diode and tunneling diode were almost the same.

Graphene-graphene Stacked Junction

Graphene-graphene junction (6) exhibited ohmic properties under a low bias voltage condition. As depicted in Fig. 2(a), a nonlinear characteristic was observed under a high bias voltage condition (>10 V). Since the density of state (DOS) of graphene linearly increased near the Dirac point, it was verified that the graphene-graphene junction was ohmic, as reported in several papers. The DOS of graphene far from the Dirac point increased non-linearly and had a divergence at van Hove singularity (7). Fig. 2(b) shows the estimated junction voltage dependence of the differential conductance. The conductance at low bias voltage was 80 µS. The differential conductance increased with an increase in the bias voltage and became approximately 8 times the initial value at an estimated junction voltage of 5 V.

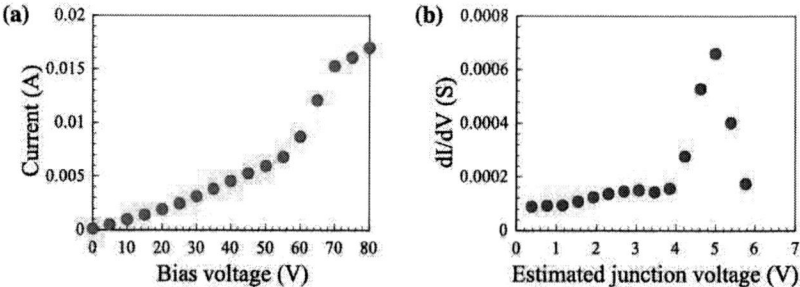

Figure 2. (a) Current-voltage characteristics of graphene-graphene junction.
(b) Differential conductance as a function of estimated junction voltage.

Infrared emission from graphene-graphene junction was expected because of the high electrical power at the high bias region. Figure 3 shows the radiation spectrum of the stacked graphene junction at 80 V, 1.3 W measured using a Fourier-transform infrared (FTIR) spectrometer. The dotted line represents the reference blackbody emission curve calculated at 300 K. A blackbody-like emission was observed. The peak wavelength of the measured infrared spectra at various power values was almost constant, 10.2 µm, which corresponds to a blackbody temperature of 284 K estimated by Planck's law. If the infrared emission from the stacked graphene junction originated from the blackbody emission, the peak wavelength should shift toward the short wavelength with an increase in input power. However, the peak wavelength of the measured FTIR spectra was almost constant with a change in electrical power. A light wavelength of 10.2 µm corresponds to a photon energy of 122 meV. The Fuchs-Kliewer (F-K) surface phonon energy of SiC (0001) is 117 meV, which is close to the measured value. The blackbody-like emission from the vertically stacked graphene junction diode should originate from the strong plasmon-phonon coupling between the graphene plasmon and the F-K phonon of the SiC substrate.

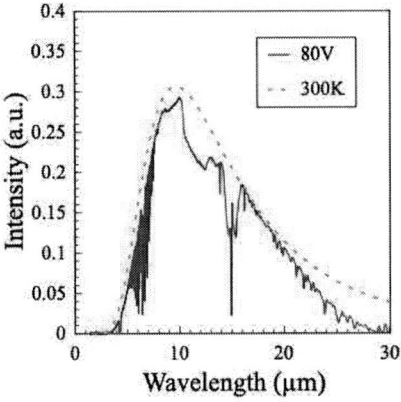

Figure 3. Radiation spectra of the stacked graphene junction measured by FTIR.

Tunneling Diode with Air Gap

As a result of the insulative layer thickness between the grapheme layers, tunneling current can flow through the barrier layer (Fig. 4(a)). Figure 4(b) shows the current-voltage (I-V) characteristics of the stacked graphene diode with an air gap insulator. A strong nonlinear I-V curve was observed. A Fowler-Nordheim (FN) plot is shown in Fig. 4(c) to revel the electrical transport mechanism. The negative linear slope of the FN plot demonstrates that the FN tunneling phenomenon was dominant in the high-electronic field region (>1 V). Tunneling electrons emitted from one graphene layer were transported across the triangular potential (Fig. 4(a)). Since controlling the thickness of the air gap was difficult, a formation of a thin dielectric layer on graphene would be the preferential choice.

Tunneling Diode with Structured Water Layer

A structured water layer with sub-1-nm-thickness was formed on the epitaxial graphene surface using DI water treatment (8). After performing DI water treatment on one of the graphene samples, the two graphene samples (both with and without the structured water layer) were bonded using the direct bonding technique. As depicted in Fig. 3(a), the structured water layer acts as a tunneling barrier. Here, the maximum tunneling current was 1 µA at +6 V. The tunneling current from the graphene layer flows through the structured water layer. The tunneling barrier thickness, estimated from the slope of the FN plot, had a value of 0.70 nm, assuming an effective electron mass ratio of 0.06. Since the electrically measured thickness coincided with the thickness measured by scanning probe microscopy, it was verified that the structured water layer acted as an insulative barrier layer.

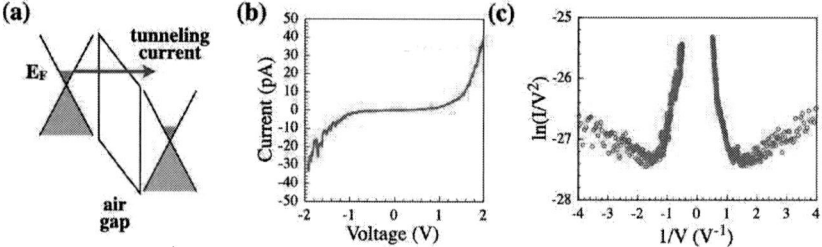

Figure 4. (a) Illustration of band diagram of FN tunneling.
(b) I-V curve of stacked graphene diode with air gap insulator.
(c) F-N plot.

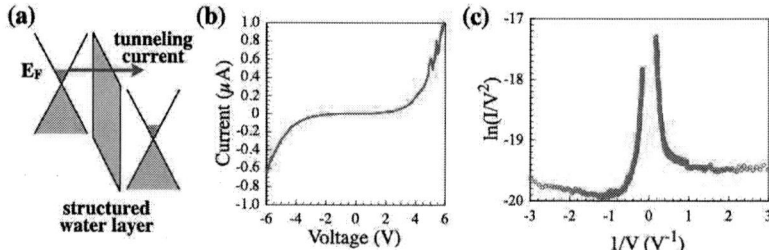

Figure 5. (a) Illustration of band diagram of FN tunneling.
(a) I-V curve of stacked graphene diode with structured water layer.
(b) F-N plot.

Conclusions

Vertically stacked graphene-graphene junction diodes were fabricated using a direct bonding technique. A high electric field was applied to the junction using single-crystal graphene on SiC substrate. A far-infrared emitter with a constant peak wavelength was realized using a new emission mechanism based on the coupling of graphene plasmon and SiC surface phonon. Tunneling diodes with an air gap and structured water layer were formed. FN tunneling characteristics were observed. Stacked graphene diodes with new functions were realized using simple device assembly processes.

Acknowledgments

This work was supported by JSPS KAKENHI Grant Number JP19H02582, JP21H01394 and the Cooperative Research Project Program of the Research Institute of Communication, Tohoku University.

References

1. K. S. Novoselov, A. K. Geim, S. V. Morozov, D. Jiang, Y. Zhang, S. V. Dubonos, I. V. Grigorieva and A. A. Firsov, *Science* **306**, 666 (2004).
2. Y. Cao, V. Fatemi, S. Fang, K. Watanabe, T. Taniguchi, E. Kaxiras and P.J.-Herrero, *Nature* **556**, 43 (2018).
3. T. Aritsuki, T. Nakashima, K. Kobayashi, Y. Ohno and M. Nagase, *Jpn. J. Appl. Phys.* **55**, 06GF03 (2016).
4. M. Nagase, H. Hibino, H. Kageshima and H. Yamaguchi, *Appl. Phys. Express* **3**, 065502 (2009).
5. M. Kitaoka, T. Nagahama, K. Nakamura, T. Aritsuki, K. Takashima, Y. Ohno, and M. Nagase, *Jpn. J. Appl. Phys.* **56**, 085102 (2017).
6. N. Murakami, Y. Sugiyama, Y. Ohno and M. Nagase, *Jpn. J. Appl. Phys.* **60**, SCCD01 (2021).
7. A. H. Castro Neto, F. Guinea, N. M. R. Peres, K. S. Novoselov, and A. K. Geim, *Rev. Mod. Phys.* **81**, 109 (2009).
8. J. Du, Y. Kimura, M. Tahara, K. Matsui, H. Teratani, Y. Ohno and M. Nagase, *Jpn. J. Appl. Phys.* **58**, SDDE01 (2019).

Single-Electron Manipulation in an Attofarad-Capacitor DRAM

K. Nishiguchi, K. Chida, and A. Fujiwara

NTT Basic Research Laboratories, Nippon Telegraph and Telephone Corp.,
Atsugi, Kanagawa 243-0198, Japan

We describe single-electron manipulation in a dynamic random
access memory (DRAM) composed of an attofarad capacitor and
nanometer-scale transistors. The motion of a single electron is
controlled and then the electron is stored in the capacitor using the
transistors, whose leakage current is suppressed to the theoretical
limit. The charge signal of the electron is amplified by another
transistor integrated in the DRAM, which functionalizes the single
electron as one bit of information for data processing. In addition to
such single-electron applications, power generation using a
conceptual analogy of Maxwell's demon detecting and manipulating
a single-electron motion is demonstrated for an understanding of the
essential correlation of data processing to its energy consumption.
These demonstrations are promising as a starting point for
constructing new technologies for the ultimate reduction in the
power consumption of data processing circuits.

Introduction

The performance of data processing circuits has been improved by miniaturizing Si field-
effect transistors (FETs). As of 2021, circuits fabricated by 2-nm process technology are
expected to be commercially available, and various trials of 1-nm process technology are
underway for the next generation of circuits. In addition, data processing circuits fabricated
by established low-cost process technologies have been used in applications where lower
performance is acceptable. These trends in semiconductor markets are expected to
accelerate with the advent of the internet-of-things (IoT) society supported by new
technologies such as the fifth-generation mobile network (5G) and artificial-intelligence
(AI). In the meantime, as the number of information-technology (IT) devices increases with
the development of the IoT society, their total energy consumption increases. This will be
a serious issue because the total energy consumption may exceed total power generated in
the world in the 2030s (1). Therefore, how to reduce the energy consumption of IT devices
can be said to be an urgent global challenge.

For sustainable development of the IoT society, we are focusing on the unique features
of single-electron devices, which control the motion of a single electron, and seeking a
basic technology for lowering the power consumption of electronic devices. One of the
most important elements in a single-electron device is its nanometer-scale structure. Many
approaches to building such a small structure involve the use of small materials such as
graphene, carbon nanotubes, and molecules. Our approach, on the other hand, takes
advantage of nanometer-scale Si FETs for their high controllability and material stability.
Using FET-based single-electron devices, we have demonstrated data processing in which

a single electron is used as one bit of information. In addition to such applications, we have studied the energy consumption of data processing from the viewpoint of thermodynamics related to Landauer's limit, which provides a physical principle pertaining to the lower theoretical limit of the energy consumption of data processing. Since our single-electron devices operate at room temperature and FETs are the most widely used active devices in the world, they can function as platforms for demonstrations of various theoretical proposals for reducing power consumption and will find practical applications in the future. In this paper, we introduce our research activities from an application-oriented perspective and fundamental physical perspective.

Principle of Si Single-Electron Devices

Fabrication

Our single-electron device is based on a DRAM fabricated on a silicon-on-insulator (SOI) wafer as shown in Fig. 1 (2). The DRAM is composed of attofarad capacitors and two FETs, hereafter referred to FET1 and FET2, which control the motion of a single electron. To utilize a single electron as one bit of information, we integrate the DRAM with another FET, hereafter referred to the amp-FET, to amplify the charge signal of a single electron in the capacitor of the DRAM.

First, nanowire channels of FET1, FET2, and the amp-FET are formed in a SOI layer. The channel width of FET1 and FET2 is 50 nm. The wire length and width of the other channel for the amp-FET are 60 and 35 nm, respectively. The two wire channels are 20 nm apart. Then, oxidation at 1000°C reduces the width of each channel to about 10 nm. Two poly-Si gates, hereafter referred to lower gates (LGs), with the gate length of 50 nm are formed on the nanowire channel of FET1 and FET2, followed by another oxidation process that reduces the gate length to about 20 nm. When both FET1 and FET2 connected in series are turned off, a single-electron box (SEB) are formed between them and a memory node (MN) is formed at the tip potion of the FET2's channel [Figs. 1(a) and (b)]. Finally, a 50-nm-thick SiO_2 interlayer and a poly-Si upper gate (UG) are formed to complete the device. Here, the UG covering the whole area shown in Fig. 1(b) is used to induce electron carriers in the SOI channels, control the current flowing through the amp-FET, and control potential energies of the SEB and MN. Figure 1(c) shows an equivalent circuit of the fabricated device. Electrons are transferred from an electron reservoir (ER) through FET1 and FET2 to the MN and detected by the amp-FET capacitively coupled to the MN.

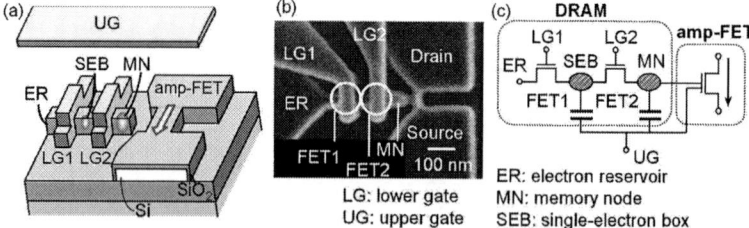

Figure 1. (a) Schematic bird's-eye view and (b) an SEM image of the single-electron device. (c) Equivalent circuit.

The fabrication method is highly compatible with conventional Si-FET fabrication and thus promises further miniaturization, high-density device integration, and highly stable characteristics, which has enabled us to develop other devices (discussed later) and will lead to single-electron applications in the future.

Principles

Single-Electron Transfer. Figure 2 shows how single-electron motion is controlled by the two FETs. When FET1 and FET2 are off, the MN and SEB are formed by energy barriers [initial state in Fig. 2(b)]. When FET1 is turned on at step (i) in Fig. 2(b), electrons in the ER enter the SEB, and then when it is turned off, single electrons are accumulated in the SEB due to the Coulomb-blockade effect [Fig. 2b(ii)] (described later). At steps (iii) and (iv) shown in Fig. 2(b), FET2 is turned on and off, whereby an electron in the SEB is transferred to the MN. These four steps compose the one electron-transfer cycle for transferring a single electron from the ER to the MN, and by repeating this cycle, electrons are transferred to the MN one after another. In the meantime, it is also possible to return electrons from the MN to the ER by lowering the potential energy of the ER.

For single-electron transfer, two technical features of high importance are provided in our devices. The first is the Coulomb blockade effect (3). When the charging energy for an electron to be injected to the SEB is larger than the thermal energy, unintentional electron injection to the SEB by thermal energy is suppressed. Accordingly, electron injection can be controlled intentionally with supply voltage. To activate the Coulomb blockade effect at room temperature, the charging energy, given by $e^2/2C$, where e is the elementary charge and C is capacitance of the SEB, must be around four times larger than the thermal energy of 26 mV, which means that the SEB's size should be 10 nm or less. Since it is not easy to achieve this size with lithography technology, electron confinement by an electric field is utilized to make the SEB smaller than the size defined by lithography. As shown in Fig. 2(b), the energy barrier formed by LG1 and LG2 spreads in the channel so that the effective planar size of the SEB is reduced (4). In addition, by applying a negative voltage to the Si substrate, electrons are confined near the upper surface of the SOI channel, which makes the SEB smaller in the longitudinal direction. These confinement effects reduce the effective size of the SEB and enable room-temperature operation of our devices.

The second technical feature for single-electron transfer is the suppression of leakage current of the FETs to the theoretical limit. The effects of subthreshold, gate, junction, and gate induction leakage increase with FET miniaturization. Though the former two can be

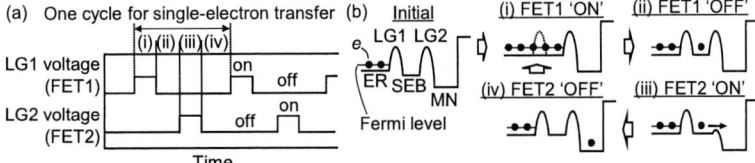

Figure 2. (a) Schematic of pulse sequences of votages applied to LG1 and LG2. The phase of LG1 voltage differs from that of LG2 voltage by π. (b) Sequences for transferring electrons from the ER to the MN through the SEB.

suppressed by improving the driving method, materials, and structures, the latter two are very difficult to be fundamentally solved because they originate from defects caused by the ion implantation used to forming the source and drain. For example, in the DRAM, since the charge in the capacitor is lost due to the leakage caused by defects, a refresh process to rewrite the data every several microseconds to several tens of microseconds is needed. To control a single electron, it is necessary to suppress these leakage currents completely. For this purpose, we use an original device structure without an impurity region. Specifically, as shown in Fig. 1, an intrinsic SOI is used as a channel, and electrons are induced at a source and drain in the SOI channel by the UG, and LG1 and LG2 control current flowing through FET1 and FET2, respectively. Consequently, leakage current caused by the defects can be eliminated, which makes leakage current suppressed to the theoretical limit determined thermally (6).

Single-Electron Detection. Figure 3(a) shows schematics of the amp-FET amplifying a charge signal originating from a single electron in the MN so that the electron can be detected. The amp-FET's channel is designed to be very close to the MN so that the channel is capacitively coupled to the MN. Thanks to this capacitive coupling, when single electrons are transferred to the MN, the current characteristics as a function of UG voltage of the amp-FET are shifted in the positive direction on the UG voltage axis. Thus, when amp-FET's current is monitored at constant UG voltage, the shift of current characteristics caused by electrons in the MN appears as a change in the current, and this current change enables the number of the electrons in the MN to be counted. For single-electron counting with higher precision, the larger shift of current characteristics caused by the electrons in the MN is managed by increasing the ratio of the capacitive coupling between the MN and the amp-FET's channel to the total capacitance of the MN and by reducing the channel size of the amp-FET (7). The capacitive coupling is increased by narrowing the distance between the detecting FET and the MN, and the total capacitance of the MN is reduced by the SiO_2 interlayer between the MN and UG. The amp-FET's channel is reduced to about 10 nm in diameter. Consequently, the charge sensitivity, meaning the minimum detectable charge, is 10^{-3} $e/Hz^{0.5}$ even at room temperature (8).

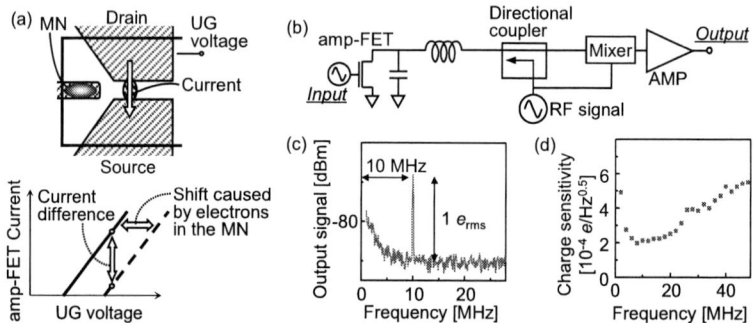

Figure 3. (a) Schematics of the structure and current characteristics of the amp-FET. (b) Equivalent circuit of a reflectometry technique for fast signal detection. (c) Power spectrum of an output signal when an input signal of 10 MHz was applied. Amplitude of input signal corresponds to a single-electron signal. (d) Charge sensitivity as a function of input-signal frequency.

On the other hand, the small amp-FET channel necessary for single-electron detection gives rise to another issue, namely difficulty of fast detection, due to a large channel resistance or RC time constant. Indeed, the sampling rate of the amp-FET is around a few tens of kilohertz. One solution for overcoming such slow operation is to use a reflectometry technique (9) [Fig. 3 (b)]. The amp-FET is connected to an inductor and capacitor, all of which compose an LCR resonator. This resonator is driven at the resonance frequency by an RF signal input via a directional coupler. When an input signal as a target signal to be detected is applied to the gate of the amp-FET, the channel resistance changes in accordance with the input signal and the resonance characteristics of the LCR resonator are modulated, and thus the reflected signal is also modulated. By mixing the reflected signal and the RF signal with a frequency mixer, the output signal shows the signal corresponding to the input signal as shown by Fig. 3(c). Although a sinusoidal wave as the input signal was used for the demonstration shown by Fig. 3(c), an arbitrary wave can also be used. In this reflectometry technique, optimization of the parameters of the resonator, such as the inductance and capacitance, makes detection speed of the amp-FET faster than the speed determined by the RC time constant. Indeed, we have succeeded in obtaining a sensitivity of around 10^{-4} $e/Hz^{0.5}$ at 20 MHz as shown in Fig. 3(d) (10).

In addition to this fast sensing, noise reduction is also possible. The frequency of the reflected signal from the LCR resonator, $f_{reflect}$, is given by $f_{RF}+f_{input}$ (or $f_{RF}-f_{input}$), where f_{RF} and f_{input} are frequencies of the RF and input signals, respectively. Therefore, it is possible to set $f_{reflect}$ higher than the frequency range in which extrinsically generated flicker noise is negligible, which leads to noise reduction. Therefore, the reflectometry improves the amp-FET's performance, including the operation speed and signal-to-noise ratio for sensing single electrons, which is useful for reading out quantum-bit, or qubit, signals for quantum computing as well as for detecting an oscillation signal of microelectromechanical systems (11).

Application Using Si Single-Electron Devices

In this section, we introduce an application that combines the single-electron transfer and detection explained in the previous section so that a single electron is used as one bit of information. All of demonstrations described here were carried out at room temperature.

Single-Electron Memory

In a conventional DRAM composed of one capacitor and one transistor, a few hundred thousand electrons are stored in the capacitor by using the transistor. This configuration is exactly the same as that of our single-electron devices explained above. Noteworthy features of our device are that a single electron is stored as one bit of information and a charge signal of the stored electron is amplified by the amp-FET next to the MN (12), which functionalizes the single-electron device as a multivalued memory operation as shown in Fig. 4. Figure 4(a) shows the change in amp-FET current when the single-electron transfer cycle shown in Fig. 2 is repeated and electrons are transferred to the MN one by one. The current decreases with a constant change in each transfer cycle, which means that the number of electrons in the MN increases by one. To prove that the electrons are transferred by the Coulomb blockade effect, the current variation of the amp-FET was evaluated when one transfer cycle was carried out by changing the ER voltage. As shown in Fig. 4(b), when ER voltage was changed, the amount of current change occurring in one transfer cycle varied stepwise, indicating that one or more electrons were transferred. When this characteristic was normalized by the quantized current change and replaced with the number of electrons and the average was taken when the measurement was repeated 100 times, a stepwise characteristic parallel to the ER voltage axis was found in the part where the average number of electrons became an integer [Fig. 4(c)]. This characteristic proves that the electron transfer is controlled by the Coulomb blockade effect. In addition, the charging energy of the SEB was deduced to be 108 meV by fitting the theoretical curve based on the Coulomb blockade effect to the experimental characteristics shown in Fig. 4(c), which also proves that single electron transfer is possible at room temperature.

When electrons are transferred by the Coulomb blockade effect as described above, the number of electrons in the MN can be controlled in accordance with the number of transfer cycles. Using this controllability of electrons in the MN, the device can be used as a multivalued memory having ten values as shown in Fig. 4(a). Since the basic driving method of our device is the same as that of conventional DRAMs, the electrons can be transferred to the MN, corresponding to steps (i), (ii), and (iii) in Fig. 2 , within 10 ns. On the other hand, our device has a long retention time of 10^4 s or more, which is much longer than that of conventional DRAMs, as shown in Fig. 5(a). This long retention time originates from ultimate suppression of current leakage from FET1 and FET2 thanks to the two-layer gate structure composed of the UG and LGs as explained above. Figure 5(b) shows the higher controllability of single-electron transfer: ER voltage can manipulate the number of electrons transferred to or ejected from the MN in one transfer cycle. In addition, the UR can also control the operating point of the Coulomb blockade effect and thus manipulate the number of the transferred electrons. These features functionalize the device as not only a digital-analog converter, in which common binary signals are converted to the number of electrons in the MN, but also as an interface between binary and single-electron signals in conventional and single-electron devices, respectively (13).

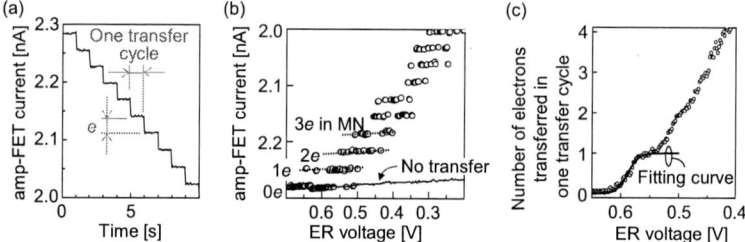

Figure 4. Single-electron transfer and detection at room temperature. (a) Change in amp-FET's current when transfer cycles are repeated. The period during which the current is constant corresponds to one transfer cycle. (b) Change in amp-FET current as a function of ER voltage after one transfer cycle. (c) Average number (open circles) of electrons transferred in the first transfer cycle as a function of ER voltage. The number of electrons are evaluated from the difference in amp-FET's current before and after one transfer cycle. The thin line is fitted to the open circles using the Coulomb blockade theory. Schematics of the structure and current characteristics of the amp-FET. (b) Equivalent circuit of the reflectometry technique for fast signal detection. (c) Power spectrum of an output signal when an input signal of 10 MHz was applied. The amplitude of the input signal corresponds to a single-electron signal. (d) Charge sensitivity as a function of input-signal frequency.

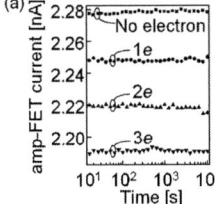 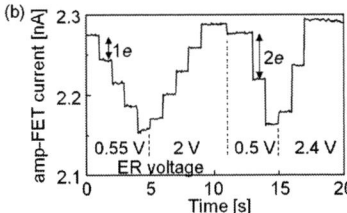

Figure 5. (a) Retention-time characteristics for various numbers of electrons in the MN. (b) Control of the number of electrons in the MN. At time indicated by broken lines, ER voltage was changed. At ER voltage of 0.55 and 2 V, one electron is injected to and ejected from the MN, respectively, in one transfer cycle. At ER voltage of 0.5 and 2.4 V, two electrons are injected and ejected, respectively.

For the single-electron-memory operation, it is necessary to carry out the electron transfer accurately by utilizing the Coulomb blockade effect. This requires the fabrication devices smaller than 10 nm, which should be possible with advanced lithography technologies. In addition, materials such as carbon nanotubes, graphene, transition-metal chalcogenide, and molecules are also promising because of their natural smallness.

Probabilistic Data Processing Using Single-Electron Shot Noise

Though the Coulomb blockade effect allows single-electron transfer with high precision, transfer with 100% accuracy is impossible because the motion of an individual electron is an essentially stochastic event. In other words, in the case of single-electron devices, the Coulomb blockade effect increases the probability that electrons will be stored in the MN but cannot fully guarantee they will be. This is an essentially unavoidable problem, and thus it is necessary to increase the charging energy to raise the transfer accuracy and to utilize circuit designs that compensate for the error with error margins, parity checks, and error correction like in conventional data processing circuits.

On the other hand, such stochastic electron motion can function as a source of physical random numbers, which enables probabilistic data processing that gives not accurate but plausible results with highly efficient time and energy. A binary bit for conventional data processing is composed of two values, '0' and '1', and they are used deterministically. A quantum bit, or qubit, used for quantum computing is composed of a superposition of '0' and '1'. In probabilistic data processing, elementary data unit, probabilistic bit (p-bit), is composed of values fluctuating in time between '1' and '0'. Although this fluctuation is not a quantum phenomenon but a classical one, e.g., based on thermal fluctuation, the concept of probabilistic computing is analogous to quantum computing (14): p-bits interact with each other. Richard. P. Feynman also mentioned the potential of probabilistic computing as a precursor to quantum computation. Recently, probabilistic data processing for factorization has been demonstrated by utilizing fluctuation of spins (15). To achieve probabilistic data processing using electronic devices, we utilize fluctuation of single-electron motion.

Figure 6(a) shows an energy band diagram along FET2's channel when single-electron motion is used for physical random numbers. Since FET1 and FET2 are on and off, respectively, the SEB is connected electrically to the ER. When the potential energy of the MN is adjusted by the UG to be lower than the Fermi energy of the ER, electrons in the ER enter the MN even when FET2 is off because the ER and MN are in an energetically non-equilibrium state. Figure 6(b) shows the change in amp-FET's current representing that electrons enter the MN one by one. It should be pointed out that the time interval at which a single electron enters the MN, δt, is always random and that a histogram of δt has an exponential distribution as shown in Fig. 6(c). In addition, the Fano factor, defined as the ratio of the average to the variance of the number of electrons entering the MN during a particular time interval, is close to one. These features mean that conduction of a single electron into the MN corresponds to shot noise, which guarantees that single-electron motion is completely random and its randomness can be used as physical random numbers. On the other hand, the average motion of electrons corresponds to current flowing through FET 2, given by $e/<\delta t>$, and thus can be controlled by LG2 voltage like the subthreshold current characteristics of an FET as shown in Fig. 6(d). In this way, the feature that individual and average motions of electrons are random and controllable, respectively, enables data processing to be probabilistically carried out.

Among such probabilistic data processing using single electrons, we introduce data processing based on a spin-glass model, which is promising for efficient data processing for image restoration and optimization problems (16, 17). In this model, data information is represented by spin up ($\uparrow$) and spin down ($\downarrow$), the total energy of the spin array is evaluated at various spin configurations, and the spin configuration giving the lowest energy corresponds to the optimum solution as the global minimum of the system as shown in Fig. 7(a). One feature of this model is that the data is processed efficiently by deriving the global minimum probabilistically. Another feature is that the total energy of a large number of spins can be considered as the sum of the energies of two adjacent spins [Fig. 7(b)], which makes it possible to simplify the system and efficiently solve optimization problems, which are difficult to solve with conventional computers based on the von-Neumann architecture. The energy of the two adjacent spins is decided by J, which originates from their interaction, and h_1 and h_2, which originate from the force applied from

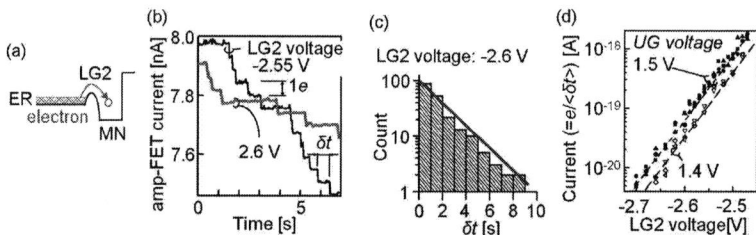

Figure 6. (a) Energy band diagram along the FET2's channel. FET1 and FET2 are on and off, respectively. (b) Electron injection from the ER to the MN. Discrete change in amp-FET current represents that a single electron enters the MN. The δt is the time interval at which each electron enters the MN. (c) Histogram of δt. The bold line is a guide for the eyes. (d) Control of δt by LG2 voltage. The y axis represents current, given by $e/\delta t$, from the ER to the MN.

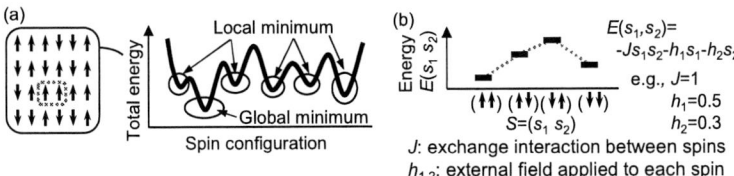

Figure 7. Conceptual diagram of spin-glass model. (a) (left) Simplified data information represented by a spin array and (right) its total energy for various spin configurations. The total energy is determined by the summation of the energy of two spins, (b). Spin pairs giving the global-minimum total energy is the optimum solution.

the outside to each of the two spins. Therefore, these three elements are parameters which decide how the data processing is implemented.

The total energy of the spin array obtained from the sum of the energies of the two spins has many local minima in addition to the global minimum, as shown in Fig. 7 (a). In order to obtain the optimum solution, it is necessary for the spin-array configuration to transit from some local minimums to the global minimum by overcoming energy barriers, and these transitions correspond to changes in the spin configuration by applying energy to spins and flipping them. Noteworthy is that the transition to the global minimum is stochastic and the efficiency for gaining the optimum solution can be increased in exchange for accuracy. For efficient operation, randomly fluctuating energy must be applied for the spin configuration to cross the energy barriers fairly. Otherwise, biased energy fluctuation increases the probability of staying at the local minimum as the solution. For this random fluctuation, we use the random electron motion shown in Fig. 6.

In this paper, data processing based on the spin-glass model is carried out by using single-electron motion instead of spins. First, as one of the most simple and most fundamental demonstrations, we derive the stable state of a spin pair $S=(s_1,s_2)$ composed of two adjacent spins—the basic element of the spin-glass model— having the energy state shown by Fig. 7(b) . The ER and LG2 voltages corresponding to spin signals s_1 and s_2 are applied so that the number of electrons injected into the MN increases as the energy of S decreases [Figs. 8(a) and (b)]. The number of injected electrons can be derived from the current difference between steps (i) and (iv). Basically, the number of the injected electrons is the maximum when S is ($\downarrow,\downarrow$) because this spin pair has the smallest energy as shown in Fig. 7(b). However, since the number of the injected electrons fluctuates due to the random single-electron motion as shown in Fig. 6, the number of the electrons at other spin pairs sometimes becomes the maximum. Using this fluctuation, the stable S is derived stochastically by repeating three steps: flipping one of the two spins; comparing the number of electrons, i.e., spin energy, before and after the spin flip; and determining the more stable S as shown in Fig. 8(c). In this paper, an ensemble average consisting of 24 spin pairs is considered to achieve more accurate and efficient operation.

Figure 9 shows the experimental results of the spin-glass operation, in which all of probabilities of each spin pair are 0.25 at first. As shown in Fig. 9(b), by repeating the process in Fig. 8(c), the probability of finding the most stable $S=(\uparrow,\uparrow)$ increases, while that

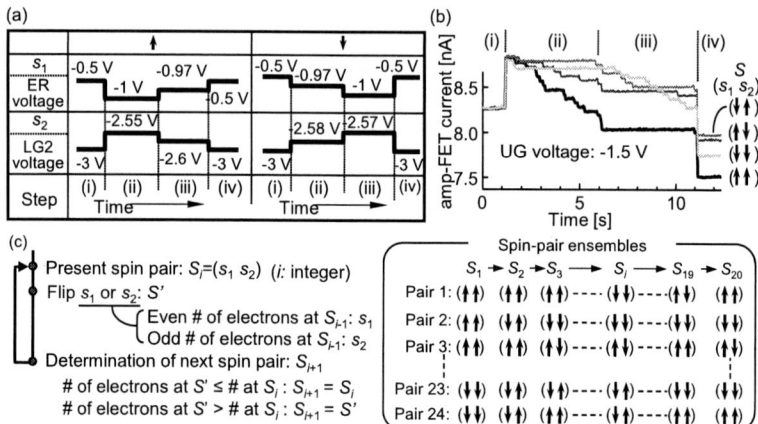

Figure 8. Schematics of the stochastic circuit. (a) Waveforms of ER and LG2 voltages to which s1 and s2 are applied, respectively. (b) Counting of electrons entering the MN for various spin pairs. Here, (i)-(iv) correspond to steps shown in (a). The abrupt increase and decrease at 1 and 11 seconds occur because the amp-FET is capacitively coupled to the ER and LG2. Since voltage conditions at (i) are the same as those at (iv), the number of electrons in the MN can be evaluated from the difference of amp-FET current between (i) and (iv). (c) Sequence to extract the most stable spin pair. The stable spin pair can be evaluated by repeating electron counting, i.e., energy comparison, at S_i and S' in the ensembles comprising 24 spin pairs.

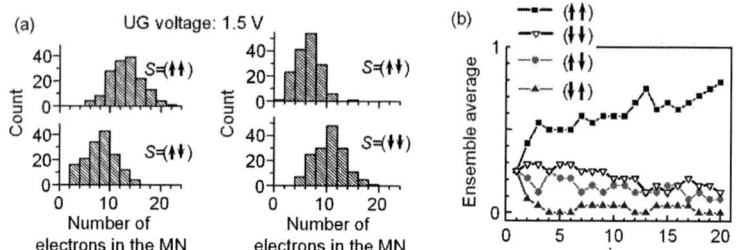

Figure 9. Stochastic evaluation of stable S. (a) Histograms of the number of electrons injected into the node when four kinds of spin pairs are applied to the circuit. (b) Change in the probability of each spin pair in the ensemble. Here, i is step of sequences shown in Fig. 7(c).

of finding the other S decreases . This is because, as shown in Fig. 9(a), the average number of injected electrons at (↑,↑) is the largest, and the number of electrons injected into the MN fluctuates at each S. This corresponds to the energy fluctuation of the spin-glass model, and it means that $(\downarrow,\downarrow)$, the local minimum state at the beginning transits to $(\uparrow,\uparrow)$ of the global minimum surmounting the energy of other spin states $[(\uparrow,\downarrow), (\downarrow,\downarrow)]$. Otherwise, without energy fluctuation, $(\downarrow,\downarrow)$ cannot transit to $(\uparrow,\uparrow)$. In addition, since the fluctuation of the number of injected electrons is completely random in this device, the most stable S can be derived without bias. Furthermore, the number of injected electrons can be controlled by UG voltage. This is equivalent to controlling the energy fluctuation, which enables control of the speed at which the state of S changes as shown by Fig. 9(b). This speed control of the state transition can improve the effect and accuracy of the data processing: initially, large energy fluctuation is applied for quick transition to the global minimum and then it is gradually reduced for stabilization at the global minimum.

An example of using the spin-glass model based on single electrons is pattern restoration as shown in Fig. 10. Figure 10(a) shows a bit map pattern in which the letter "N" is collapsed, and the aim is to restore it with the spin-glass model. First, white and black pixels constituting a bit map pattern are replaced with $\uparrow$ and $\downarrow$ spins, respectively. Then the total energy between spins constituting the pattern is evaluated from $\sum E(s_1,s_2)$, where $\sum$ represents the sum of all the adjacent spin pairs. The goal is to explore the spin configuration of the minimum total energy by flipping the spins in the above manner. Consequently, the collapsed "N" is restored as shown in Fig. 10(b).

The pattern restoration introduced here becomes possible by using random motion of single electrons as physical random numbers, and this can also be used for other applications such as pattern matching (18) and high-quality imaging based on stochastic resonance (19). Such stochastic or probabilistic data processing could enable efficient processing in time and energy in comparison with conventional deterministic data processing. Indeed, in a demonstration using stochastic resonance, which is known as a mechanism supporting efficient activity in biosystems, including humans, it was confirmed that the intrinsic energy necessary for the data processing of one bit can be reduced to the order of 10^{-19} J. Although probabilistic data processing can be achieved by using other electric random number generators, the use of single-electron motion guarantees high-quality randomness as physical random numbers and enables the function of a random number generation to be integrated with logic circuits, such as NAND and NOR. Therefore, probabilistic data processing can be said to be one of the applications taking advantage of single electrons.

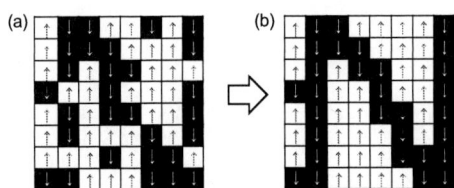

Figure 10. Pattern restoration using the spin-glass model. A bit-map pattern of "N" (a) before and (b) after the restoration.

Approach to Ultimate Low Power Consumption for Data Processing

In the above chapter, we introduced single-electronic devices aiming for data processing with low power consumption. However, the power consumption does not simply decrease even if the number of electrons representing one bit is reduced. In this chapter, we discuss power consumption from the fundamental viewpoint.

Landauer's Limit

Landauer's limit, proposed by Rolf Landauer in 1961, is a physical principle giving the lower limit of energy consumption of computation. The acquisition of one bit of information means that entropy, or Shannon entropy, decreases by $k\ln2$ (k: Boltzmann's constant), which means that energy corresponding to thermal energy of $kT\ln2$ (T: absolute temperature) related to the reduction in the entropy must be applied from the viewpoint of thermodynamics (20). For our demonstration introduced later, we give another explanation based on the case of a memory storing bit information of '0' or '1'. Figure 11(a) shows a potential diagram showing the memory storing '0'. This information can be held by providing an energy barrier between '0' and '1'. In order to prevent '0' from becoming '1' due to thermal fluctuation or noise, the height of the energy barrier must be equal to or higher than the thermal energy of $kT\ln 2$. On the other hand, to rewrite the information of the memory from '0' to '1', energy larger than the energy-barrier height is applied, and then the state of the memory is shifted to '1' as shown in Fig. 11(b). Therefore, the lower limit of the energy holding and writing information accurately is derived as $kT\ln2$, i.e., Landauer's limit. If accuracy of the information is not necessary, it could be possible to reduce these energies. However, when there is no accuracy, the information is meaningless. Therefore, the lower limit of energy consumption of computation is derived as $kT\ln2$. This idea is universal for any device utilizing electrons, spins, and so on.

Landauer's limit is very fundamental and far from the energy consumed by actual electric devices. Therefore, it had seemed to be investigated only theoretically. However, now, with the advance to the IoT society in the future, it is expected that the power consumption of IT devices will exceed power generation in the world (1). Moreover, the Semiconductor Industry Association of America (SIA) proposed in 2015 that the power consumption of IT devices should be reduced by two or three orders of magnitude from the current level. This proposed power consumption corresponds to only three or four orders of magnitude larger than Landauer's limit. Therefore, now is the time to consider Landauer's limit seriously. Until now, the discussion of Landauer's limit has been limited to theoretical considerations, but in recent years, demonstrations of it have been reported

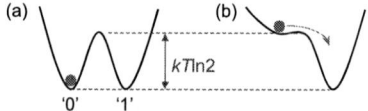

Figure 11. Schematic diagram of the potential of the memory that holds the bit information ('0' or '1'). The circles indicate the state of the memory. When (a) '0' memory information is rewritten to (b) '1', it is necessary to exceed the potential between '0' and '1'. When rewriting memory information from (a) '0' to (b) '1', it is necessary to exceed the potential between '0' and '1'.

(21, 22), and the environment for a deep understanding of Landauer's limit has been established.

Maxwell's Demon.

Landauer's limit provides the fundamental relationship between "Information" and "Energy" and has a strong implication of "Energy" for dealing with "Information". On the other hand, a more substantive idea linking "Information" and "Energy" is Maxwell's demon, which is a thought experiment proposed by James C. Maxwell (Fig. 12). When there are cold and hot molecules in two rooms partitioned by a door, these molecules distribute randomly due to thermal motion and are monitored by Maxwell's demon in real time. When hot molecules in the left room move to the right room, the demon opens and closes the door and confines the hot molecules in the right room. The door is kept closed when cold molecules move from left to right. In the meantime, when the cold molecule moves from the right room to the left room, the door is opened and closed, and the cold molecules are confined in the left room. Repeating such operation has hot and cold molecules confined in the right and left rooms, respectively, and generates a temperature difference between them. This temperature difference is an energy source: if the door between the right and left rooms could slide along the two rooms, it could work to the outside due to the temperature difference, which is so-called a Szilard engine. Noteworthy is that the molecules move randomly due to thermal energy and the demon generates energy without touching the molecules. Since this energy generation by the demon seems to violate the second law of thermodynamics, the demon has been a paradox in the thermodynamics field for a long time. After various discussions of this paradox (23-27), information thermodynamics provided an important concept (28, 29). In the theory of the information thermodynamics, when the demon observes a molecule and gets its "Information", energy corresponding to mutual information, a quantity of information representing the accuracy of actual and observed information, is consumed. Then, this energy is transferred to the molecule by the demon opening and closing the door. Consequently, the second law of thermodynamics is satisfied. Landauer's limit can be explained only under the condition that the bit information of '0' and '1' are generated with the same probability, and a problem is that Landauer's limit is not thermodynamically valid when the probability for '1' is different from that for '0'. Information thermodynamics can solve such a problem of Landauer's limit and, more interestingly, it provides a new information insight, namely that "Information" is "Energy". Therefore, information thermodynamics has attracted much attention and shown promise for supporting the development of future circuits that enable ultimately efficient information processing based on, for instance, the unique mechanism of efficient biosystems (30). In addition, recent demonstrations of information thermodynamics, including Maxwell's demon, by using microbeads, electrons, and photons have also contributed to increaseing activity in this field of study (31-33).

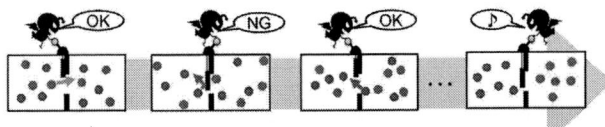

Figure 12. Conceptual diagram of Maxwell's demon. The demon opens and closes the door according to the movement of warm and cold molecules.

Power Generation by Maxwell's Demon. One of the fundamental concepts of information thermodynamics is Maxwell's demon. Though a few demonstrations of Maxwell's demon using electronic devices, i.e., electrons, have been reported, they were carried out at low temperatures. To expand the concepts of information thermodynamics to actual electronic circuits, room-temperature demonstrations are needed. In the previous chapter, we explained how single-electron detection and transfer are realized at room temperature by using a DRAM structure, which means that these functions are applicable to the demonstration of Maxwell's demon.

Here, using FETs, electric power generation is demonstrated with the analogy of Maxwell's demon (34). Figure 13 shows schematics of the demonstration. The device structure shown in Fig. 13(b) is almost the same as that shown in Fig. 1. The ER and MN shown in Fig. 1 correspond to the entrance and exit shown in Fig. 13, respectively, and electrons can flow out through the exit, whose potential energy can be controlled as a parameter explained later. FET1 and FET2 are connected in series and the SEB is formed electrically between them. The amp-FET functions as the demon monitoring electrons in the SEB. For simplicity, in Fig. 13(a), FETs are represented by doors. When the left door, i.e., FET1, is open, electrons move randomly between the entrance and SEB by thermal energy. When the electron enters the SEB, the demon closes the left door and confines the electron in it. Then, the demon opens the right door, i.e., FET2, and closes the door when the electron leaves the SEB through the door. These steps, composed of monitoring electrons and consequently opening/closing the doors, transfer the electron from the entrance to the exit with a higher energy state, which corresponds to power generation.

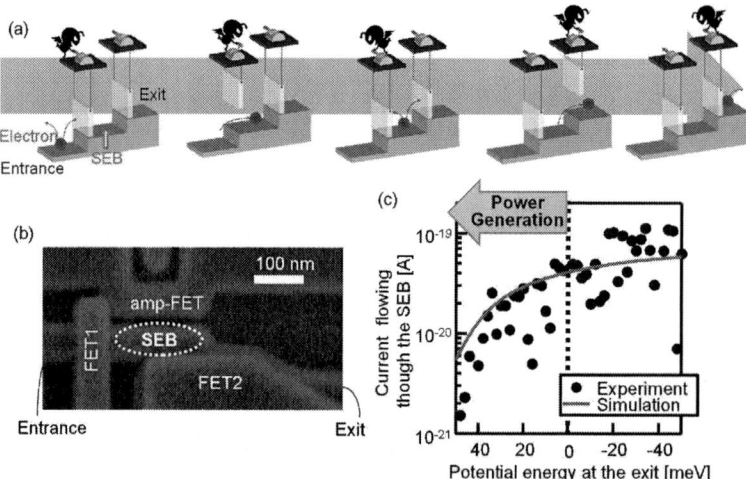

Figure 13. (a) Schematic view and (b) an scanning electron microscope image of the devices for demonstrating Maxwell's demon. (b) Rectification characteristics of noise using Maxwell's demon. The potential at the exit shows the height of the potential relative to the entrance. The current is derived by multiplying the number of electrons flowing from the inlet to the outlet in one second by the charge (1.6×10^{-19} C).

Figure 13(c) shows the characteristics of current flowing through the SEB when potential energy of the exit is changed. When the potential energy is negative, electrons flow in a forward direction to the exit, whose energy potential is lower than that of the entrance. On the other hand, even when the potential energy is negative, current flows, which means that electrons are transferred to the higher potential energy, i.e., power generation, thanks to the operation based on Maxwell's demon. Since this demonstration is achieved by using FETs, we can expect not only other demonstrations of proposals for low power consumption of data processing from the viewpoint of information thermodynamics but also applications inspired from such demonstrations in the future.

Summary

By manipulating and detecting single electrons in a DRAM composed of nanometer-scale FETs, a data processing circuit utilizing a single electron as one bit of information was demonstrated. While deterministic operation is possible by using the Coulomb-blockade effect, probabilistic data processing is also possible by taking advantage of the randomness of single-electron motion. In addition to such applications, Maxwell's demon was demonstrated from the viewpoint of information thermodynamics, which may provide hints to establishing new data processing circuits with ultralow power consumption. Since the device is composed of nanometer-scale FETs for high stability and controllability, it can be used as an optimum platform for carrying out demonstrations of some theoretical proposals. Since a small structure is essential for realizing the present device and others like it, FETs fabricated with future advanced fabrication technology as well as from other small materials such as graphene, carbon nanotubes, and molecules are promising for further progress in this field.

Acknowledgments

We thank Y. Takahashi of Hokkaido University, and H. Inokawa, and Y. Ono of Shizuoka University for helpful discussions.

References

1. *Impact of Progress of Information Society on Energy Consumption (Vol. 1): Current Status and Future Prospects for Power Consumption of IT Equipment*, Center for Low Carbon Society Strategy, Japan Science and Technology Agency (2019).
2. K. Nishiguchi, H. Inokawa, Y. Ono, A. Fujiwara, and Y. Takahashi, *Electron. Lett.* **40**, 229 (2004).
3. K. K. Likharev, *Proc. IEEE*, **87**, 606 (1999).
4. K. Nishiguchi, H. Inokawa, Y. Ono, A. Fujiwara, and Y. Takahashi, *Appl. Phys. Lett.* **85**, 1277 (2004).
5. K. Nishiguchi, C. Olivier, H. Namatsu, S. Horiguchi, Y. Ono, A. Fujiwara, Y. Takahashi, and H. Inokawa, *Jpn. J. Appl. Phys.* **44**, 7717 (2005).
6. K. Nishiguchi, A. Fujiwara, Y. Ono, H. Inokawa, and Y. Takahashi, *IEEE Electron Dev. Lett.* **28**, 48 (2007).
7. M. H. Devoret and R. J. Schoelkopf, *Nature* **406**, 1039 (2000).

8. K. Nishiguchi, C. Koechlin, Y. Ono, A. Fujiwara, H. Inokawa, and H. Yamaguchi, *Jpn. J. Appl. Phys.* **47**, 8305 (2005).
9. R. J. Schoelkopf, P. Wahlgren, A. A. Kozhevnikov, P. Delsing, and D. E. Prober, *Science* **280**, 1238 (1998).
10. K. Nishiguchi, H. Yamaguchi, A. Fujiwara, H. S. J. Zant, and G. A. Steele, *Appl. Phys. Lett.* **103**, 143102 (2013).
11. K. Nishiguchi and A. Fujiwara, *22nd International Microprocesses and Nanotechnology Conference (MNC2020)*, Hiroshima, Japan (Oct. 2019).
12. K. Nishiguchi, H. Inokawa, Y. Ono, A. Fujiwara, and Y. Takahashi, *Appl. Phys. Lett.* **85**, 1277 (2004).
13. K. Nishiguchi, A. Fujiwara, Y. Ono, H. Inokawa, and Y. Takahashi, *Appl. Phys. Lett.* **88**, 183101 (2006).
14. J. Jordan, M. A. Petrovici, O. Breitwieser, J. Schemmel, K. Meier, M. Diesmann, and T. Tetzlaff, *Sci. Rep.* **9**, 18303 (2019).
15. W. A. Borders, A. Z. Pervaiz, S. Fukami, K. Y. Camsari, H. Ohno, and S. Datta, *Nature* **573**, 390 (2019)
16. H. Nishimori, *Statistical physics of spin glasses and information processing*, Oxford University Press, Oxford (2001).
17. K. Nishiguchi and A. Fujiwara, *International Electron Devices Meeting (IEDM)*, Washington, USA (Dec. 2007).
18. K. Nishiguchi and A. Fujiwara, *Nanotechnology* **20**, 175201 (2009).
19. K. Nishiguchi and A. Fujiwara, *Jpn. J. Appl. Phys.* **50**, 06GF04 (2011).
20. R. Landauer, IBM J. Research and Develop. 5, 183 (1961).
21. A. Berut, A. Arakelyan, A. Petrosyan, S. Ciliberto, R. Dillenshneider, and E. Lutz, *Nature* **483**, 187 (2012).
22. A. O. Orlov, C. S. Lent, C. C. Thorpe, G. P. Boechler, and G. L. Snider, *Jpn. J. Appl. Phys.* **51**, 06FE10 (2012).
23. L. Szilard, *Zeitschrift für Physik* **53**, 840(1929).
24. L. Brillouin, *J. Appl. Phys.* **22**, 334 (1951).
25. C. H. Bennett, *IBM J. Res. Dev.* **17**, 525 (1973).
26. C. H. Bennett, *Int. J. Theoret. Phys.* **21**, 905 (1982).
27. R. Landauer, *IBM J. Res. Dev.* 5, **183** (1961).
28. T. Sagawa and M. Ueda, *arXiv*: 1111.5769v2 (2012).
29. J. M. R. Parrondo, J. Horowitz, and T. Sagawa, *Nat. Phys.* **11**, 131 (2015).
30. S. Ito and T. Sagawa, *Nat. Commun.* **6**, 7498 (2015).
31. S. Toyabe, Y. Sagawa, M. Ueda, E. Muneyuki, and M. Sano, *Nat. Phys.* **6**, 988 (2010).
32. J. V. Koski, V. F. Maisi, T. Sagawa, and J. P. Pekola, *Phys. Rev. Lett.* **113**, 030601 (2014).
33. M. D. Vidrighin, O. Dahlsten, M. Barbieri, M. S. Kim, V. Vedral, and I. A. Walmsley, *Phys. Rev. Lett.* **116**, 050401 (2016).
34. K. Chida, S. Desai, K. Nishiguchi, and A. Fujiwara, *Nat. Commun.* **17**, 464 (2018).

Chapter 3

Novel Integration

(Invited) Multi-Channel AlGaN/GaN Power Rectifiers: Breakthrough Performance up to 10 kV

Y. Zhang[a*], M. Xiao[a], Y. Ma[a], Z. Du[b], H. Wang[b], A. Xie[c], E. Beam[c], Y. Cao[c], and K. Cheng[d]

[a] Center for Power Electronics Systems, Virginia Polytechnic Institute and State University, Virginia 24060, USA
[b] Ming Hsieh Department of Electrical and Computer Engineering, University of Southern California, California 90089, USA
[c] Qorvo Inc., Texas 75081, USA
[d] Enkris Semiconductor Inc., Jiangsu 215123, China
*E-mail: yhzhang@vt.edu

> High-voltage power rectifiers are widely used in renewable energy processing, electric grids, industrial motor drives, pulse power systems, among other applications. Today's high-voltage rectifier market is dominated by bipolar Si diodes up to 6.5 kV, which suffer from slow reverse recovery. Wide-bandgap SiC unipolar diodes have been pre-commercialized up to 10 kV, which allows for a much higher switching speed. Recently, we have developed a new generation of high-voltage rectifiers based on the multi-channel AlGaN/GaN platform, which highlight a series of novel device designs incorporating the stacked two-dimensional electron gas (2DEG) channels, p-n junctions, and 3-D fin structures. With these innovations, the performances of our unipolar 1.2-10 kV multi-channel AlGaN/GaN Schottky rectifiers well exceed the Si and SiC 1-D limit, at the same time possessing a lower cost as compared to SiC counterparts. This paper reviews our efforts in the design, fabrication and characterization of these GaN devices. Our results show the tremendous promise of GaN for medium-voltage and high-voltage power electronics applications.

Introduction

Power semiconductor devices providing low on-resistance, high switching speed, and high blocking voltage are central to improving the efficiency of electrical energy processing in electric vehicles, data centers, electric grids, among other applications. High-voltage (HV, 1.7 kV – 10 kV) power rectifiers are ubiquitously used in electricity grid, renewable energy processing, industrial motors, and electrified transportation (Fig. 1). Today's HV rectifier market is dominated by bipolar Si p-n junction diodes up to 6.5 kV. However, they have a very slow switching speed due to poor reverse recovery. A superior alternative that allows fast switching is the SiC Schottky barrier diode (SBD) or junction Schottky barrier (JBS) diode. SiC JBS diodes up to 10 kV have been pre-commercialized by Cree/Wolfspeed (1) and used in R&D power electronics applications (2-4). However, the epitaxial and fabrication costs of HV SiC rectifiers are much higher than Si counterparts, which hinder their wide adoption and commercialization.

Another material that offers superior physical properties for power devices is GaN. Compared to Si and SiC, GaN has a higher critical electric field (E-field) and a unique high-mobility channel, the two-dimensional electron gas (2DEG). After two decades of development, lateral GaN power devices have been commercialized up to 650 V as a superior replacement of the similarly-rated Si devices (5,6). However, the continued voltage and power upscaling of lateral GaN devices have encountered great challenges, many of which stem from the limited current capability of the thin 2DEG channel, crowded E-field near the device surface, and the resulted difficulties in thermal management (7,8). Despite the reports of high-voltage lateral GaN SBDs up to 9 kV (9,10), their differential on-resistance is significantly larger than that of SiC SBDs. This situation leads to a common belief that the vertical GaN structure is more favorable for HV power devices. However, despite many high-performance vertical GaN devices at 1.2-2 kV classes (11-13), the highest breakdown voltage (*BV*) reported in vertical GaN devices is only 5 kV (14).

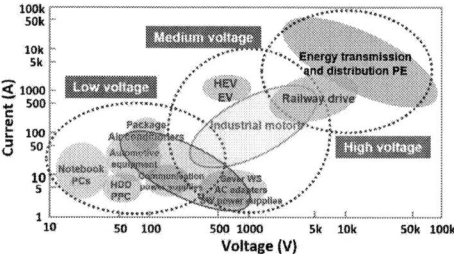

Figure 1. Representative applications of low-, medium- and high-voltage power rectifiers.

Recently, our team has proposed the multi-channel lateral AlGaN/GaN platform for HV power devices and demonstrated a series of novel lateral GaN rectifiers that feature stacked 2DEG channels and innovative device designs. The fundamental rational to pursue multi-channel AlGa/GaN devices is that they retain a high 2DEG mobility while concurrently leveraging the benefits of vertical devices, e.g., spatially-distributed electron current and E-field, to maximize the power density. Our devices are fabricated on 4-inch AlGaN/GaN-on-sapphire wafers that host five stacked 2DEG channels and have a sheet resistance (R_{SH}) below 120 Ω/sq, which is at least 4~5-fold lower than that in commercial lateral GaN devices. The cost of this GaN-on-sapphire wafer is estimated to be 2~3-fold lower than a SiC wafer (6). However, voltage upscaling in multi-channel devices is more challenging than that in the single-channel counterpart, due to the excess charges and the resulted leakage current and E-field crowding. To overcome these challenges, we have demonstrated various device innovations, including the p-GaN edge termination (15), 3-D junction-fin anode (16), and p-GaN reduced surface field (RESURF) structure (17). With these innovations, the performance of our multi-channel AlGaN/GaN rectifiers up to 10 kV has well exhibited the unipolar 1-D limit of SiC devices.

Large-Diameter Multi-Channel Wafer

Power device design aims at concurrent realization of lower on-resistance (R_{ON}) and high *BV*. By stacking multiple 2DEG channels, the 2DEG density can be increased proportionally, therefore reducing the wafer R_{SH} and device R_{ON}. AlGaN/GaN multi-channel epitaxy has been initially demonstrated around the 2010s by Molecular Beam Epitaxy (MBE) (18). However, MBE is usually not suitable for large-diameter, high-volume wafer production. Recently, 4-inch multi-channel wafers have become available by Metal-Organic Vapour-Phase Epitaxy (MOCVD) on various substrates including Si, SiC, sapphire and GaN. The 4-inch, 5-channel, GaN-on-sapphire wafer produced by Enkris Semiconductor Inc. (Fig. 2) possesses a 2DEG density of 3.7×10^{13} cm^{-2}, a 2DEG mobility of 1475 cm^2/V·s, and an R_{SH} of 110 Ω/sq, with the R_{SH} being over 3-fold lower than the usual value of a single-channel wafer. A p-GaN cap layer can be continuously grown on the multi-channel AlGaN/GaN structure (17). Even with the p-GaN depletion, the multi-channel wafer retains a 2DEG density of 1.75×10^{13} cm^{-2}, a 2DEG mobility of 2010 cm^2/V·s, and an R_{SH} of 178 Ω/sq. This low R_{SH} is key to enabling a device R_{ON} much lower than that of similarly-rated SiC and Si rectifiers.

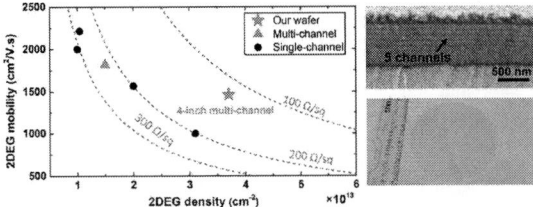

Figure 2. Comparison the 2DEG properties of our wafer with other reports. Scanning electron microscopy image showing 5 channels, and a photo of the large-diameter wafer.

P-GaN Termination: E-field Management

The excess charges from multiple 2DEG channels often induce E-field crowding, and this issue is particularly crucial for Schottky rectifiers, as their BV is typically limited by the peak E-field at the Schottky contact region. Proper edge termination is thus essential to alleviate the E-field crowding or potentially move the peak E-field away from the Schottky contact, just like the basic principle of the JBS design (19,20). For lateral AlGaN/GaN SBDs, field plate is a widely-used edge termination structure (Fig. 3(a)). However, its effectiveness requires precise control over the field plate geometry, such as dielectric thickness and field plate length. Additionally, the complex interfaces between dielectrics and semiconductors often result in device instability under high E-field, leading to preliminary breakdown.

To address the above challenges, we developed a new termination structure using a p-GaN layer epitaxially grown on AlGaN/GaN (Fig. 3(b)) (15). Owing to the vertical depletion enabled by the p-n junction, the E-field lines spreads out, and their distribution becomes more uniform. The peak electric field is also moved from the Schottky contact to the edge of p-GaN, thereby shielding the Schottky contact from high E-field. Compared to the field plate, this p-GaN termination possesses a wide design window in terms of doping concentration and thickness and comprises few dielectric interfaces. Its fabrication is fully compatible with today's foundry process for manufacturing the p-gate normally-off high-electron-mobility transistors (HEMTs), thereby opening the possibilities for integrating high-voltage rectifiers with GaN power ICs. This p-GaN termination structure enables the demonstration of the first 3.3-kV AlGaN/GaN multi-channel SBDs exceeding the SiC unipolar limit (15).

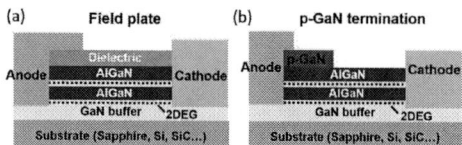

Figure 3. Schematic of (a) field plate and (b) p-GaN termination for multi-channel AlGaN/GaN SBDs.

P-GaN RESURF: Towards the Multi-Channel Super-Junction

Superjunction is one of the most successful concepts in the history of power device development, which relies on alternative n- and p-doped pillars and can break the theoretical trade-off between R_{ON} and BV of 1-D drift regions (21). In GaN, the vertical superjunction like that in Si and SiC has not been experimentally demonstrated (22,23). An alternative superjunction in lateral GaN relying on the natural balance in polarization charges, which is referred to as "polarization superjunction" or "natural superjunction", has been experimentally demonstrated on single- or double-channels (10,24). However, the reported performance of GaN polarization superjunction devices is not significantly better than the 1-D counterparts, and the R_{ON} vs. BV trade-off is inferior to SiC devices.

In an undoped multi-channel structure, according to the ideal band theories, a balance in polarization charges is expected to be naturally established in each channel, thereby forming a multi-channel polarization superjunction. If this is true, a p-GaN termination would be sufficient for E-field management, and the E-field distribution in the access region should be quite uniform benefited from the superjunction properties. However, the net charge in our experimental multi-channel devices was found to be non-zero, and the additional donors are present. To reach the overall charge balance in this "unbalanced superjunction", we proposed a novel design for multi-channel devices, the p-GaN RESURF structure (Fig. 4(a)). Compared to the p-GaN termination, the p-GaN RESURF layer extends to the region nearing the cathode, and its total acceptor charges balance the net donor charges in the multi-channel structure when the device is blocking voltage (Fig. 4(b)) (17). In the fabrication, C-V measurements on a test structure along with the p-GaN etch can identify the critical p-GaN thickness to reach the overall charge balance (17).

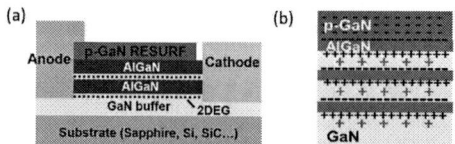

Figure 4. Schematic of the (a) RESURF multi-channel AlGaN/GaN SBD and (b) charge balance in this RESURF multi-channel structure.[17]

Fig. 5(a) and (b) show the forward characteristics of the p-GaN-terminated SBDs with an anode-to-cathode distance (L_{AC}) of 98-148 μm and the p-GaN RESURF SBDs with L_{AC} of 48-123 μm. A same turn-on voltage (V_{on}) was extracted to be 0.6 V for all SBDs A low on-resistance (R_{on}) of 28.7 and 31 Ω·mm was extracted in the p-GaN

RESURF SBDs with 98 and 123 μm L_{AC}, respectively. Fig. 5(c) shows reverse I-V characteristics of the three types of SBDs with increased L_{AC}. With an identical L_{AC} (e.g., 98 μm), the p-GaN RESURF SBD shows a BV about 2-fold higher than that of the SBD without edge termination and about 1.5-fold higher than that of the SBD with p-GaN termination. Despite the non-uniform E-field, an average lateral E-field ($E_{AVE} = BV / L_{AC}$) can be calculated, which is useful for the lateral device design. At a BV of ~5 kV and above, the E_{AVE} of non-terminated SBDs, p-GaN-terminated SBDs, and p-GaN RESURF SBDs are 0.42-0.47 MV/cm, 0.59-0.64 MV/cm, 0.94-1 MV/cm, respectively. The p-GaN RESURF SBD with 98 μm L_{AC} shows a BV of 9.15 kV; the device with 123 μm L_{AC} was measured to 10 kV (our measurement limit) repeatedly without showing any degradation.

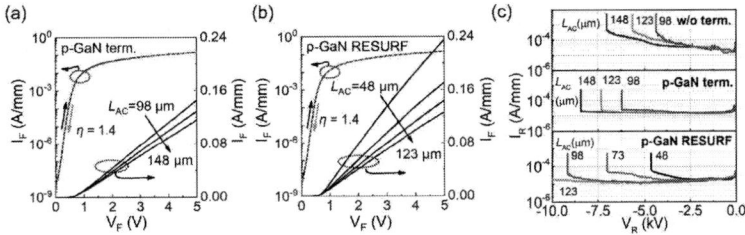

Figure 5. Forward I-V characteristics of multi-channel AlGaN/GaN SBDs with various L_{AC}, one with (a) p-GaN termination and (b) the other with p-GaN RESURF structures. (c) Reverse I-V characteristics at various L_{AC} for multi-channel SBDs with and without terminations as well as with the ones with p-GaN termination or p-GaN RESURF.

The SBD with a 98-μm L_{AC} shows a BV of 9.15 kV and a specific R_{ON} of 29.5 mΩ·cm², rendering a Baliga's figure of merit (FOM=BV^2/R_{ON}) of 2.84 GW/cm². The SBD with a 123-μm L_{AC} shows a BV over 10 kV and a R_{ON} of 39 mΩ·cm², which is 2.5-fold lower than the R_{ON} of the state-of-the-art 10-kV SiC JBS diodes. The Baliga's FOMs of our 4.6-10 kV GaN SBDs well exceed the SiC unipolar limit.

Junction-Fin Anode: Minimizing the Leakage Current

The leakage current reduction is another challenge facing high-voltage multi-channel devices, as the Schottky contact to each AlGaN/GaN channel may suffer from barrier lowering effect subject to high bias. Addressing this challenge is our junction-fin-anode, a three-dimensional anode structure that comprises p-n junctions wrapping around the multi-2DEG-fins (Fig. 6(a)) (16). Compared to the planar p-GaN termination or RESURF structure, the 3-D wrapped p-n junctions can provide a stronger depletion of the 2DEG channel. When the device is reverse biased, the junction-fin assists the Schottky contact for charge depletion and shields the Schottky contact from seeing high biases.

Our design can be illustrated by the equivalent circuit model of the entire rectifier (Fig. 6(b)) This model includes an equivalent series connection for a sidewall SBD, a junction-fin-gated HEMT, and a p-gate HEMT. As the reverse bias increases, the sidewall SBD is pinched off, and then the two HEMTs. The voltage drop on the sidewall SBD is clamped at the threshold voltage of the junction-fin-gated HEMT, which is merely a few volts. This clamping occurs regardless of the reverse bias at the cathode, which can reach thousands of volts. Operating in this manner, the leakage current of the entire rectifier is equal to that of one of the sidewall SBDs biased at a few volts.

In our prototyped 5-kV device, we realized the junction-fin structure by regrowth of p-GaN on top of the fin, and the addition of a p-type nickel oxide at the fin sidewalls (16). The resulting rectifier delivers a BV up to 5.2 kV, and when operating at 90% of this BV, the leakage current is just 1.4 µA/mm. The specific R_{ON} is 13.5 mΩ·cm², rendering a power figure-of-merit (FOM) exceeding the SiC unipolar limit. Subsequently, large-area multi-channel AlGaN/GaN SBDs with junction-fin anodes have been fabricated. They are capable of handling a 1.5 A current, have a leakage current measured in microamps, and a total charge of 13 nC (Fig. 6(c)-(d)). As compared with commercial and R&D SiC SBDs with similar voltage and current ratings, our multi-channel GaN SBDs exhibit a significantly lower forward voltage and charges.

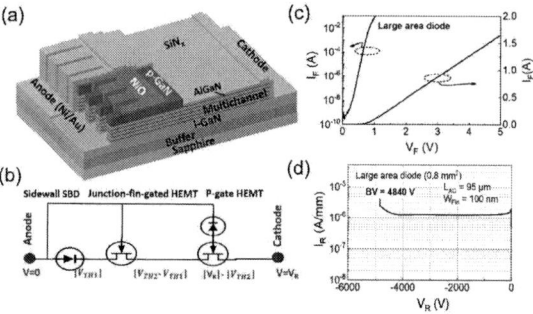

Figure 6. (a) 3-D schematic and (b) equivalent circuit model of the multi-channel AlGaN/GaN SBDs with the junction-fin anode. (c) Forward and (d) reverse I-V characteristics of the fabricated large-area devices.

We note that the junction-fin anode structure can be not only compatible to the p-GaN termination structure as we demonstrated in (16) but also the RESURF structure. To showcase this viability, we scrutinized the functions of the p-GaN termination and junction-fin anode. As shown in Fig. 7, the p-GaN termination determines the device BV, while the junction-fin-anode reduces the leakage current. This suggests, by combining the p-GaN RESURF and junction-fin anode in multi-channel devices, a further reduction of leakage current can be readily envisioned in 10-kV+ multi-channel AlGaN/GaN SBDs.

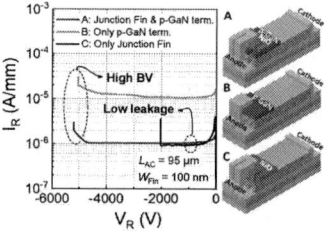

Figure 7. Reverse I-V characteristics of devices with only p-GaN termination, only junction-fin, and both structures.

Benchmark and Summary

Fig. 8 benchmarks the specific R_{ON} vs. BV of our multi-channel AlGaN/GaN SBDs with the state-of-the-art GaN SBDs, SiC JBS/SBDs, and Ga_2O_3 SBDs with a BV over 2 kV. A contact finger length of 3 μm (25) was added to $L_{AC}+L_A$ in $R_{ON,SP}$ calculation. Our 1.2-10 kV multi-channel SBDs show a Baliga's FOM (BV^2/R_{ON}) of over 2.8 GW/cm², which is the highest among all reported multi-kilovolts SBDs and well exceeds the 1-D SiC unipolar limit. The practical performance limit of AlGaN/GaN multi-channel devices $[R_{on,sp} = BV^2/(q\mu_{2DEG}n_{2DEG}E_{AVE}^2)]$ was found to reach the vertical GaN limit using $E_{AVE} = 1$ MV/cm and $R_{SH} = 150$ Ω/sq.

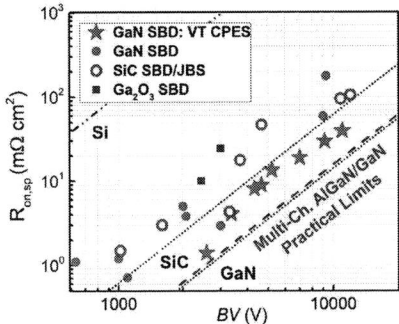

Figure 8. The differential R_{ON} vs. BV benchmark for our SBDs and the state-of-the-art GaN, SiC, and Ga_2O_3 HV SBDs. The Si, SiC, GaN bulk limits and the multi-channel lateral AlGaN/GaN practical limit are also plotted.

In particular, our multi-channel RESURF SBD with a 123-μm L_{AC} shows a BV over 10 kV and a R_{ON} of 39 mΩ·cm², which is 2.5-fold lower than the R_{ON} of the state-of-the-art 10-kV SiC JBS diodes. In addition, our 10-kV GaN SBD has a V_{ON} (0.6 V) lower than that of 10-kV SiC JBS diode (>1 V (26)), suggesting a lower forward voltage (V_F). Assuming an 80 mA/mm forward current, the switching FOM ($V_F \cdot Q_C$) of a 10-kV, 0.3-A GaN multi-channel RESURF SBD is projected to be 15.7 nC·V, which is even lower than that of a commercial 3.3-kV, 0.3-A SiC SBD (30.8 nC·V, GAP3SLT33-214 GeneSiC Semiconductor) (no Q_C data available for higher-voltage SiC SBDs).

These superior device performances, in addition to the lower wafer and fabrication cost of lateral GaN devices, all show the great potential of pushing GaN power devices into the HV realm and the good promise of the multi-channel AlGaN/GaN devices as the platform technology for HV GaN devices.

Acknowledgments

The Virginia Tech authors acknowledge the partial support by the National Science Foundation under Grant ECCS-2036740 and the Power Management Industry Consortium of the Center for Power Electronics Systems, Virginia Tech.

References

1. J. B. Casady, V. Pala, D.J. Lichtenwalner, E.V. Brunt, B. Hull, G. Wang, J. Richmond, S.T. Allen, D. Grider, and J.W. Palmour, in *Proc. PCIM Eur. 2015 Int. Exhib. Conf. Power Electron. Intell. Motion Renew. Energy Energy Manag.* (2015), pp. 1–8.
2. J. Wang, T. Zhao, J. Li, A.Q. Huang, R. Callanan, F. Husna, and A. Agarwal, *IEEE Trans. Electron Devices* **55**, 1798 (2008).
3. R. Zhang, X. Lin, J. Liu, S. Mocevic, D. Dong, and Y. Zhang, *IEEE Trans. Power Electron.* **36**, 2033 (2021).
4. R. Zhang, X. Lin, J. Liu, S. Mocevic, D. Dong, and Y. Zhang, in *2020 32nd Int. Symp. Power Semicond. Devices ICs ISPSD* (2020), pp. 246–249.
5. Y. Zhang, A. Zubair, Z. Liu, M. Xiao, J.A. Perozek, Y. Ma, and T. Palacios, *Semicond. Sci. Technol.* **36**, 054001 (2021).
6. Y. Zhang, A. Dadgar, and T. Palacios, *J. Phys. Appl. Phys.* **51**, 273001 (2018).
7. Y. Zhang, M. Sun, Z. Liu, D. Piedra, H.S. Lee, F. Gao, T. Fujishima, and T. Palacios, *IEEE Trans. Electron Devices* **60**, 2224 (2013).
8. Y. Zhang and T. Palacios, *IEEE Trans. Electron Devices* **67**, 3960 (2020).
9. A. Colón, E.A. Douglas, A.J. Pope, B.A. Klein, C.A. Stephenson, M.S. Van Heukelom, A. Tauke-Pedretti, and A.G. Baca, *Solid-State Electron.* **151**, 47 (2019).
10. H. Ishida, D. Shibata, H. Matsuo, M. Yanagihara, Y. Uemoto, T. Ueda, T. Tanaka, and D. Ueda, in *2008 IEEE Int. Electron Devices Meet. IEDM* (2008), pp. 1–4.
11. Y. Zhang, M. Sun, J. Perozek, Z. Liu, A. Zubair, D. Piedra, N. Chowdhury, X. Gao, K. Shepard, and T. Palacios, *IEEE Electron Device Lett.* **40**, 75 (2019).
12. J. Liu, M. Xiao, Y. Zhang, S. Pidaparthi, H. Cui, A. Edwards, L. Baubutr, W. Meier, C. Coles, and C. Drowley, in *2020 IEEE Int. Electron Devices Meet. IEDM* (2020), p. 23.2.1-23.2.4.
13. J. Liu, M. Xiao, R. Zhang, S. Pidaparthi, H. Cui, A. Edwards, M. Craven, L. Baubutr, C. Drowley, and Y. Zhang, *IEEE Trans. Electron Devices* **68**, 2025 (2021).
14. H. Ohta, K. Hayashi, F. Horikiri, M. Yoshino, T. Nakamura, and T. Mishima, *Jpn. J. Appl. Phys.* **57**, 04FG09 (2018).
15. M. Xiao, Y. Ma, K. Cheng, K. Liu, A. Xie, E. Beam, Y. Cao, and Y. Zhang, *IEEE Electron Device Lett.* **41**, 1177 (2020).
16. M. Xiao, Y. Ma, Z. Du, X. Yan, R. Zhang, K. Cheng, K. Liu, A. Xie, E. Beam, Y. Cao, H. Wang, and Y. Zhang, in *2020 IEEE Int. Electron Devices Meet. IEDM* (2020), p. 5.4.1-5.4.4.
17. M. Xiao, Y. Ma, K. Liu, K. Cheng, and Y. Zhang, *IEEE Electron Device Lett.* **42**, 808 (2021).
18. Y. Cao, K. Wang, G. Li, T. Kosel, H. Xing, and D. Jena, *J. Cryst. Growth* **323**, 529 (2011).
19. Y. Zhang, Z. Liu, M.J. Tadjer, M. Sun, D. Piedra, C. Hatem, T.J. Anderson, L.E. Luna, A. Nath, A.D. Koehler, H. Okumura, J. Hu, X. Zhang, X. Gao, B.N. Feigelson, K.D. Hobart, and T. Palacios, *IEEE Electron Device Lett.* **38**, 1097 (2017).

20. Y. Zhang, M. Sun, Z. Liu, D. Piedra, M. Pan, X. Gao, Y. Lin, A. Zubair, L. Yu, and T. Palacios, in *2016 IEEE Int. Electron Devices Meet. IEDM* (2016), p. 10.2.1-10.2.4.
21. F. Udrea, G. Deboy, and T. Fujihira, *IEEE Trans. Electron Devices* **64**, 720 (2017).
22. M. Xiao, R. Zhang, D. Dong, H. Wang, and Y. Zhang, *IEEE J. Emerg. Sel. Top. Power Electron.* **7**, 1475 (2019).
23. Y. Ma, M. Xiao, R. Zhang, H. Wang, and Y. Zhang, *IEEE J. Electron Devices Soc.* **8**, 42 (2020).
24. H. Ishida, D. Shibata, M. Yanagihara, Y. Uemoto, H. Matsuo, T. Ueda, T. Tanaka, and D. Ueda, *IEEE Electron Device Lett.* **29**, 1087 (2008).
25. R. Zhang, J.P. Kozak, M. Xiao, J. Liu, and Y. Zhang, *IEEE Trans. Power Electron.* **35**, 13409 (2020).
26. J. Lynch, N. Yun, and W. Sung, in *2019 31st Int. Symp. Power Semicond. Devices ICs ISPSD* (2019), pp. 223–226.

Chapter 4

Metrogy and Characterization

62

Millisecond Annealing by Atmospheric Pressure Thermal Plasma Jet and Direct Imaging of Temperature Distribution using Optical Interference Contactless Thermometry (OICT)

S. Higashi, K. Matsuguchi, T. Sato, and H. Hanafusa

Graduate School of Advanced Science and Engineering, Hiroshima University, Higashihiroshima, Hiroshima 739-8530, Japan

> Ultra-rapid thermal annealing in millisecond has been performed by atmospheric-pressure thermal plasma jet (TPJ) and temperature measurement based on optical interference contactless thermometry (OICT) is investigated. On the basis of optical interference in silicon wafer induced by the change in refractive index and its analysis, transient temperature variation is obtained with millisecond time resolution. Moreover, OICT imaging allow us to obtain the temperature information in planar and depth direction simultaneously.

Introduction

Ultra-rapid thermal annealing (URTA) in millisecond is one of the key process technologies in semiconductor device fabrication. Various heat sources such as laser, flash lamp, and plasma are utilized to perform instantaneous heat treatments. Temperature measurement of the wafer with a high time resolution is indispensable not only to achieve a precise temperature control and a reproducibility of the annealing process itself, but to control the uniformity of device performances. Infrared radiation thermometers have been proposed to measure silicon wafer temperature (1), however, plasma processing involves various difficulties such as plasma radiation (2). In previous work, the present authors proposed an optical-interference contactless thermometry (OICT) for interferometric temperature measurements with high temporal resolution and the ability to measure a substrate surface temperature from the backside has been demonstrated (3-5). In this work, the OICT is improved with accurate values of the thermo-optic coefficient (TOC) of silicon and more-realistic simulations. The TOC of silicon is measured precisely from room temperature to 800 K. Transient temperature of a silicon wafer during a thermal plasma jet (TPJ) annealing is measured by the OICT and a thermocouple simultaneously to evaluate the absolute temperature accuracy and response speed of the OICT. The wafer is heated with the TPJ under a wide range of annealing speed from 10 to 10^5 K/s. Moreover, the OICT has been extended to imaging the temperature distribution inside the wafer.

Experimental

In the present work, a probe laser (λ = 1310 nm, 14 mW) was used to irradiate silicon wafers, and the transmissivity or reflectivity were measured. The refractive index of the wafer changes with temperature, and the interference condition changes.

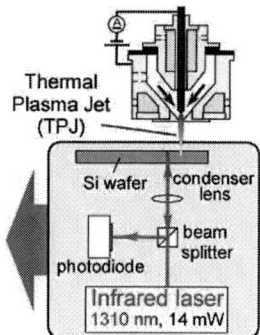

Fig. 1. Experimental set up for OICT. URTA was performed by TPJ irradiation with scanning silicon wafers in the front.

Consequently, the transmissivity or reflectivity oscillate with temperature. The temperature is derived by analyzing the transient change in transmissivity or reflectivity with a known TOC and optical path of the substrate. Here, the TOC is obtained from an *ex-situ* measurement, as shown later. A double-sided polished 8–12-Ωcm-resistivity and 0.525-mm-thick p-type silicon (100) wafers are used in the following experiments. In the experiment, URTA was performed by atmospheric pressure TPJ irradiation to silicon wafers (see Fig. 1). TPJ was generated by DC arc discharge of Ar gas with flow rate (f) of 1.0 to 3.0 L/min, DC power (P) in the range of 0.48–1.79 kW. TPJ was generated by blowing the arc plasma out through 0.6 or 2-mm-diameter orifices. A 1.5 cm × 1.5 cm silicon wafer was moved linearly with the optics by a motion stage in front of the TPJ with a scanning speed (v) in the range of 10–1500 mm/s, and the distance between the plasma source and the wafer (d) was varied in the range of 0.5–30 mm to control the wafer temperature. The measured waveforms of reflectivity are reproduced by heat diffusion simulation and optical interference analysis, the details of which are described elsewhere (3).

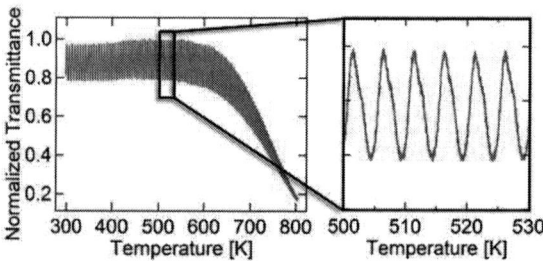

Fig. 2. Transmittance of silicon wafer measured at 1310 nm as a function of temperature. Clear oscillation is observed due to interference of probe laser light and change in refractive index.

TOC measurement

A vacuum chamber is evacuated to 1 Pa, and a 2 cm × 2 cm silicon wafer is fixed in the center and heated gradually at 5 K/min by a plate heater while maintaining thermal equilibrium, and the transmissivity and wafer temperature are measured by a 0.05-mm-diameter type-R thermocouple (TC). Figure 2 shows an example of the oscillation waveform of normalized transmissivity as a function of the wafer temperature. Clear oscillation was observed in the transmissivity waveform ant the amplitude began to decrease from 600 K, and the oscillation almost disappeared at 800 K. This is caused by the band to band absorption increase (6). Figure 3 shows the refractive index of silicon at a wavelength of 1310 nm extracted from the peak-valley of the oscillation waveform. The second-order approximate equation is given by

$$n_{Si} = 3.46 + 1.18 \times 10^{-4} \times T + 1.30 \times 10^{-7} \times T^2,$$ [1]

where T is the Kelvin temperature. The measured value is comparable to the reported value from room temperature to 600 K (7).

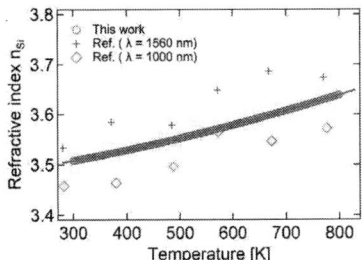

Fig. 3. Thermo optic coefficient (TOC) of silicon extracted from Fig. 2.

Temperature measurement during URTA

Transient reflectivity was measured during URTA of silicon wafer performed under TPJ scan v of 1500 mm/s. The red lines in Fig. 4 are the measured transient reflectivity, with the region in each dashed rectangle being enlarged and shown in the next graph. A clear oscillation is observed in the time range of ~10 s down to the microsecond region as shown in Fig. 4. The yellow arrow indicates the phase inversion point of the oscillation, at which heating ends and heat diffuses through the wafer results in cooling. The number of oscillations is the same before and after this arrow because the wafer is heated from room temperature and returns there again. We performed simulation to reproduce the measured waveform as shown by the dotted blue lines in Fig. 4. It is possible to fit the phase of oscillation quite accurately in a wide time range from microseconds to 10 s. Figure 5 shows transient variation of wafer temperature during URTA obtained from OICT. The surface temperature reached ~480 K, while backside temperature was ~400K at maximum. It should be noted that the OICT gives transient temperature change in arbitrary position inside the wafer.

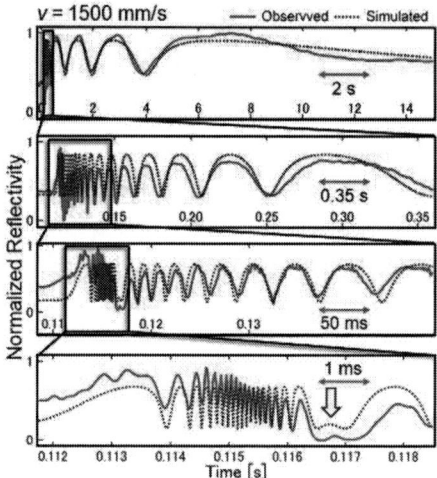

Fig. 4. Transient reflectivity observed during TPJ irradiation (red line) and simulated result (blue line). Experimental results are well reproduced by the simulation in the time range from 10s to ms.

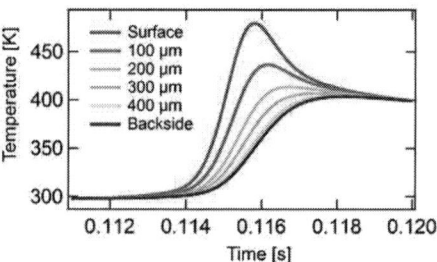

Fig. 5. Transient variation of temperature in silicon wafer. Based on OICT, temperature changes in different depth are obtained with millisecond time resolution

OICT imaging

The OICT was extended to 2-dimensional observation of transient reflectivity by the introduction of a high-speed camera (HSC). In order to capture interference fringes in 2-dimension, the photodiode shown in Fig. 1 was replaced by a HSC. Probe laser light was expanded by a pair of lenses to illuminate the area where the HSC observed. The other measurement and analyses are basically the same. Figure 6(a) shows a snapshot of the observed transient reflectivity image taken under TPJ irradiation condition of $P = 1.18$ kW, $f = 1.0$ L/min, $v = 400$ mm/s, and $d = 2$mm, with 0.6 mm orifice. Clear fringes are observed due to the temperature variation and resulting distribution of refractive index inside the wafer. Experimental observation was reproduced by OICT analysis as shown in Fig. 6(b). From the present analysis on reflectivity images, temperature distributions in the wafer was obtained as shown in Fig. 7. Maximum temperature reached ~460 K near the center of TPJ irradiation point. Since the URTA was performed in millisecond, temperature gradient in the depth direction appears as shown in Fig. 7(b). As understood from these results, a capture of fringe image is enough to obtain the temperature distribution in OICT imaging. This new approach is quite useful to obtain the temperature instantaneously.

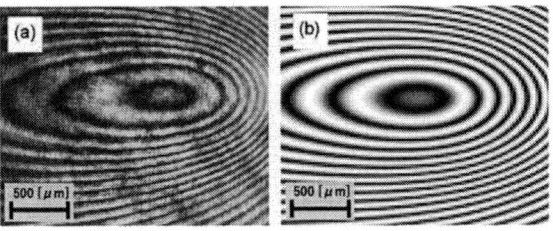

Fig. 6. Captured image during TPJ irradiation to silicon wafer (a) and calculated fringes based on OICT analysis model.

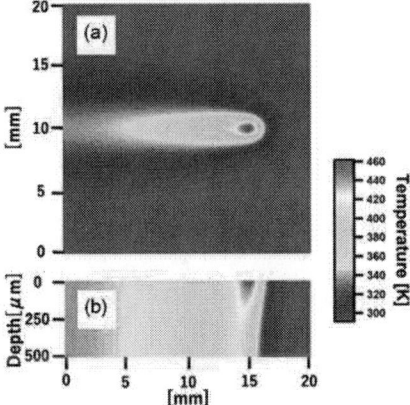

Fig. 7. Temperature distributions in silicon wafer during TPJ annealing obtained by OICT. Planar temperature distribution (a) and in-depth distribution (b) are obtained based on the analysis shown in Fig. 6.

Conclusions

Measurement of silicon wafer temperature during URTA in millisecond has been successfully performed based on OICT. Transient temperature variation from ~10s to 1 ms has been precisely measured. By the extension of the present technique to OICT imaging, temperature information in 3.5-dimension (x, y, z, and time) is instantaneously obtained. This will be a strong tool for temperature control in URTA in semiconductor processing.

Acknowledgments

A part of this work was supported by Adaptable and Seamless Technology transfer Program through Target-driven R&D (A-STEP) from Japan Science and Technology Agency (JST) Grant Number JPMJTR20RS.

References

1. T. Arai, H. Fujita, and K. Oguri, Thin Solid Films **165**, 139 (1988).
2. T. Shimada, T. Miura, W. Xie, T. Yanase, and T. Nagahama, Measurement **102**, 244 (2017).
3. S. Higashi , H. Kaku, T. Okada, H. Murakami, and S. Miyazaki, Jpn. J. Appl. Phys. **45**, 4313 (2006).
4. T. Okada, S. Higashi, H. Kaku, N. Koba, H. Murakami, and S. Miyazaki, Jpn. J. Appl. Phys. **45**, 4355 (2006).
5. T. Okada, S. Higashi, N. Koba, H. Kaku, H. Murakami, and S. Miyazaki, Thin Solid Films **515**, 4897 (2007).
6. J. C. Sturm and C. M. Reaves, IEEE Trans. Electron Devices 39, 81 (1992).
7. J. A. McCaulley, V. M. Donnelly, M. Vernon, and I. Taha, Phys. Rev. B **49**, 7408 (1994).

Two-Dimensional Characterization of Wide-Bandgap Materials and Contact
Interfaces by Using Scanning Internal Photoemission Microscopy

K. Shiojima

Graduate School of Electrical and Electronics Engineering, University of Fukui, 3-9-1
Bunkyo, Fukui, Japan

> Scanning internal photoemission spectroscopy has been developed
> to map the electrical characteristics of metal/semiconductor
> interfaces nondestructively. We conducted two-dimensional
> characterization of wide-bandgap Schottky contacts such as GaN,
> SiC, and oxide semiconductors. Our experimental demonstrations
> of the mapping characterization are reviewed from the aspects of
> (A) thermal degradation, (B) device degradation by applying high-
> voltage, (C) process-induced surface damages, (D) grain boundaries
> of semiconductors and printed metal particles, and (E) expansion to
> semiconductor/semiconductor and metal-insulator-semiconductor
> interfaces. This technique was confirmed to be useful for the
> development of the wide-bandgap-semiconductor high-power
> devices.

Introduction

Metal/semiconductor (M/S) interfaces are one of the most important components in the
device fabrication. In the conventional characterization, such as current-voltage (I-V) and
capacitance-voltage methods, the average values of the contacts are used. However, the
actual M/S interfaces are not uniform. There are various origins of inhomogeneities, such
as crystal defects, surface contamination layers, interfacial reaction, and metal grain
boundaries.

Under these circumstances, T. Okumura and K. Shiojima invented the original
method, termed scanning internal photoemission microscopy (SIPM) to map the electrical
characteristics of M/S interfaces nondestructively in 1989 (1,2). In the early days, we
demonstrated mapping for Si and GaAs Schottky contacts using infrared lasers as a light
source (3,4). Recently, as wide-bandgap semiconductor materials have been intensively
studied for high-power RF and switching electron devices, we reconstructed SIPM for
these by using visible lasers. We demonstrated two-dimensional (2-D) characterization of
interfacial reaction and surface damages in SiC, GaN, and Oxide-semiconductor Schottky
contacts (5-18). We also demonstrated that SIPM is available for
semiconductor/semiconductor (S/S) hetero-interfaces and metal-insulator-semiconductor
(MIS) interfaces (19,20). In this paper, we review our experimental results on SIPM
mapping so far from the aspects of (A) thermal degradation, (B) device degradation by
applying high-voltage, (C) process-induced surface damages, (D) grain boundaries of
semiconductors and printed metal particles, and (E) S/S and MIS interfaces.

SIPM Method

When a monochromatic light with a photon energy (hv) greater than Schottky barrier height ($q\phi_B$) is incident on the metal/semiconductor interface, carriers in the metal can surmount the Schottky barrier and a photocurrent may be generated as shown in Figure 1(a). This is known as the internal photoemission effect. The $q\phi_B$ can be determined from the measured photocurrent, using Fowler's equation (21), as follows (Figure 1(b)):

$$Y^{1/2} \propto (hv - q\phi_B) \qquad [1]$$

where Y is the photoyield which is the photocurrent per incident photon number. When hv is close to the fundamental absorption edge, due to the generation of electron-hole pairs, a large photocurrent flows, such as in a solar cell. In the SIPM measurements, by focusing and scanning the laser beam over the interface, we obtained a 2-D image of Y as shown in Figure 1(c). The beam spot diameter of used lasers is less than 2 μm in a visible light region. We typically used red ($\lambda = 660$ nm, $hv = 1.88$ eV), green ($\lambda = 517$ nm, $hv = 2.40$ eV), and violet ($\lambda = 405$ nm, $hv = 3.06$ eV) lasers in our SIPM measurements, and obtained 2-D images of Y. Using Y maps at each hv, we obtained an image of $q\phi_B$ according to Eq. [1].

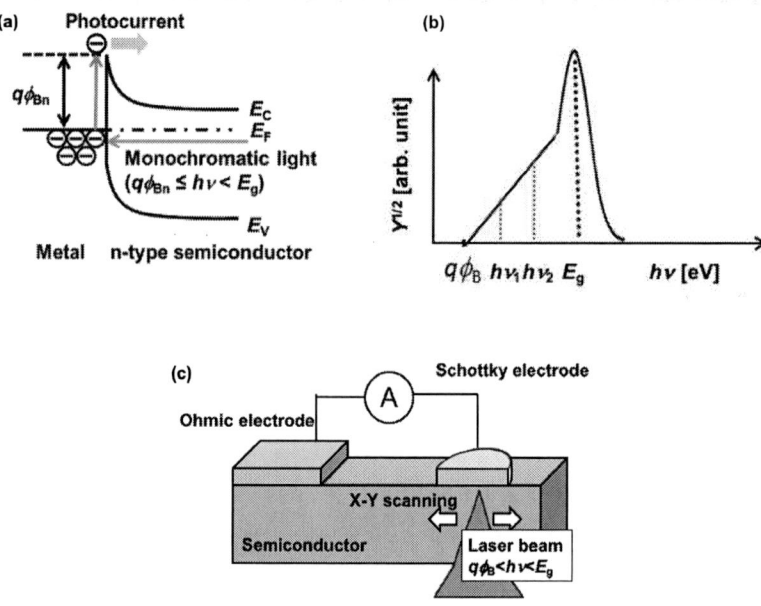

Figure 1. (a) Energy band diagram of a metal/n-semiconductor interface, (b) internal photoemission spectrum, and (c) device structure and SIPM set-up.

Demonstrations of SIPM

A: Thermal Degradation

One of the most important issues in the device reliability is thermal stability. In most cases, the interfacial reaction between metal and semiconductor is responsible for the thermal degradation of the Schottky contacts. In the early days of the development of SIPM, we conducted an annealing study of Au/Pt/Ti/n-GaAs Schottky contacts (4). This structure is used for the gate electrodes of GaAs metal-semiconductor field-effect transistors. After annealing at 400° C, Pt atoms diffused to the interface and $PtAs_2$ was formed. However, we confirmed by SIPM that the uniformity was preserved. After annealing at 480-500° C, the I-V characteristics degraded, and large Y areas appeared at the periphery of the electrode in the SIPM results, because Au atoms migrated to the interface and a β–AuGa alloy providing low $q\phi_B$ was formed.

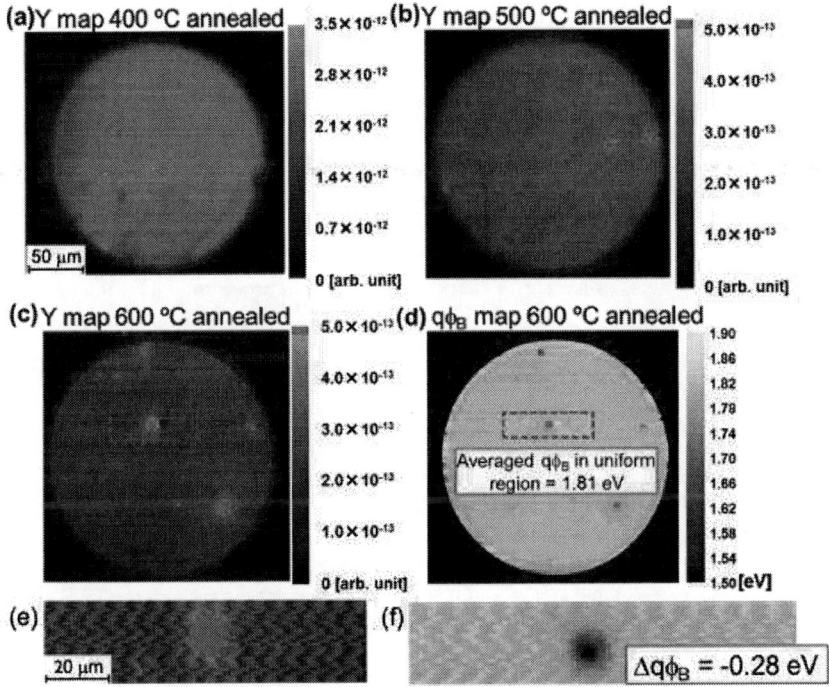

Figure 2. Thermal degradation of the Ni/n-SiC Schottky contacts. Y images after annealing at (a) 400, (b) 500, (c) 600° C, and (d) a $q\phi_B$ images. Magnified (e) Y and (f) $q\phi_B$ images of the dotted rectangular regions. After Reference 5, Copyright 2015, Japan Society of Applied Physics.

For the development of high-power devices, the thermal degradation is more important. In the case of Ni/n-SiC Schottky contacts as shown in Figure 2, the local interfacial reaction between Ni and SiC was observed in the Y image after annealing at 500° C or higher (5,6). The lower $q\phi_B$ was determined in such regions. In the case of Au/Ni/n-GaN Schottky contacts, we found that surface scratches on the metal dot enhanced interfacial degradation. On the other hand, for the Cu/Ti/α-Ga$_2$O$_3$ Schottky contacts as shown in Figure 3, there was an improvement in uniformity of the Schottlky characteristics after annealing. (7). Ti can easily react with oxygen; hence, the uniform interfacial reaction proceeded. For the Au/Ti/Pt/α-Ga$_2$O$_3$, Y increased in the contact periphery. The local interfacial reaction to form PtGa with a low $q\phi_B$ occurred. For the Cu/Ti/Fe/α-Ga$_2$O$_3$, the degradation with the small Y was extended from the edge in a line shape. The significant reaction between Fe and oxygen caused the gap formation at the interface.

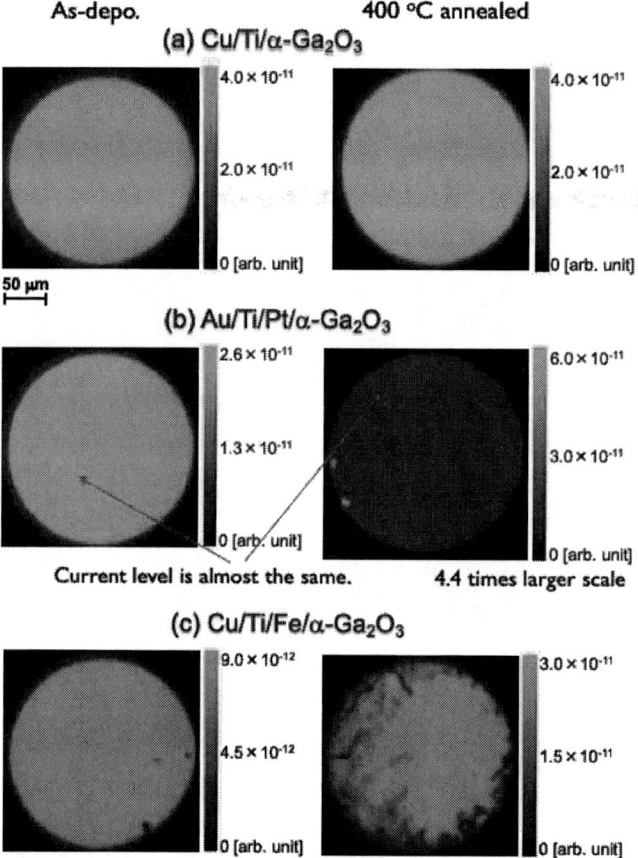

Figure 3. Y images of (a) Cu/Ti, (b) Au/Ti/Pt, and (c) CuTi/Fe contacts formed on α-Ga$_2$O$_3$ before and after annealing at 400° C. After Reference 7, Copyright 2019, John Wiley & Sons, Inc.

B: Degradation by Applying Voltage

Another advantage of the wide-bandgap semiconductors for the high-power device applications is a large breakdown voltage. After applying a reverse voltage stress of 30 V to the Au/amorphous In–Ga–Zn–O Schottky contacts, the *I-V* characteristics became leaky and a local degradation at the edge of the contacts was clearly visualized in the *Y* image, while no symptoms of the degradation were observed in the conventional optical microscope image (Figure 4) (8).

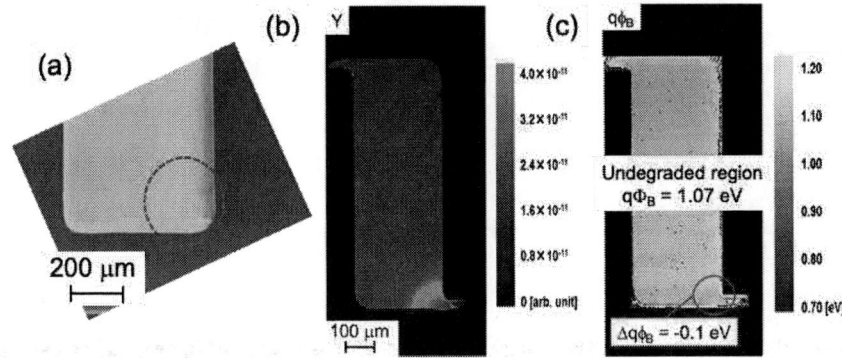

Figure 4. (a) Microscope, *Y*, and $q\phi_B$ images of the Au/In–Ga–Zn–O Schottky contacts after applying a reverse voltage stress of 30 V. After Reference 8, Copyright 2017, John Wiley & Sons, Inc.

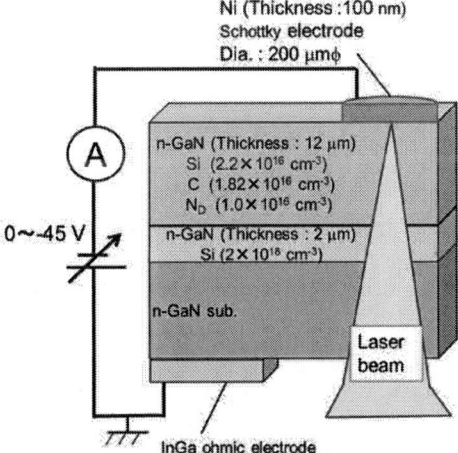

Figure 5. Vertical Ni/n-GaN Schottky contacts formed on a thick n-GaN drift layer grown on a freestanding GaN substrate with a voltage-applied SIPM configuration.

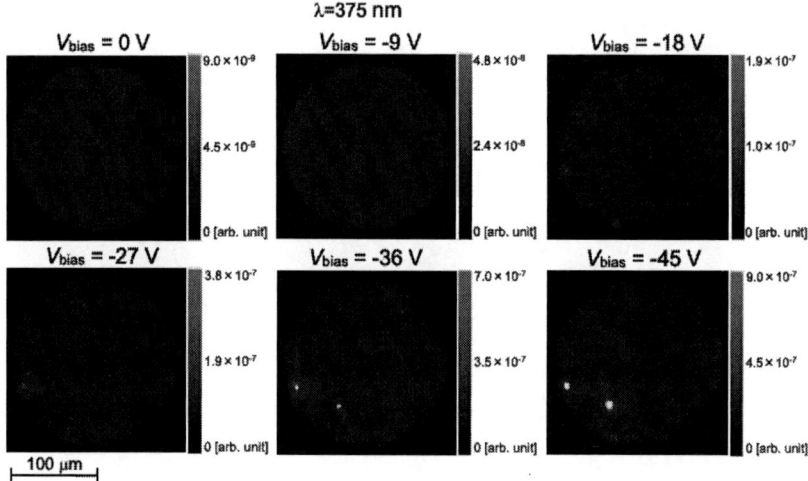

Figure 6. Y images under the near-UV irradiation of the Ni/n-GaN Schottky contact with the applying voltages from 0 to -45 V. After Reference 9, Copyright 2019, Japan Society of Applied Physics.

Applying reverse bias voltage (V_{bias}) down to -45 V during the Y measurement is possible in our SIPM (9). For most of the Ni/n-GaN Schottky contacts formed on a thick n-GaN layer grown on a freestanding GaN substrate, uniform distribution of Y was observed over the electrode. On the other hand, for the contacts with a slightly larger reverse current, the Y distribution was also uniform at $V_{bias} = 0$ V, but over $V_{bias} = -36$ V, Y increased intensively at small spots as shown in Figure 6. After the SIPM measurements, the I–V characteristics became leaky, and the same spots were observed in the microscope image. These results indicate that SIPM is useful for in-situ monitoring of the initial stage of the degradation under applying reverse bias voltage.

C: Surface Damages

The electrical characteristics of the Schottky contacts are a sensitive nature to the damage located on the semiconductor surface. We applied SIPM to the damage characterization. Firstly, the surface damage was intentionally induced by focused ion beam, or selective ion-implantation: Ga ions to n-GaAs, N ions to n-GaN, and N ions to n-SiC (3,10,11). The implanted images were clearly observed as the same Y images, and no extension of the induced damage was found (Figure 7). For the GaAs and SiC contacts, Y increased in the implanted regions, while for the GaN, Y decreased. The different roles of the damage as generation or recombination centers were clarified.

We also demonstrated SIPM for more moderate surface damage induced by inductive coupled plasma (ICP) etching (12). For Pd Schottky electrodes on n-GaN surfaces including selectively ICP-etched regions, we could clearly visualize the etched regions in the Y map, where Y increased 1.4 times and $q\phi_B$ decreased by 0.32 eV, as compared with unetched regions as shown in Figure 8. Upon annealing at 700 °C and 800 °C, both Y and $q\phi_B$ values recovered. Additionally, SIPM successfully characterized selectively neutral-beam-etched damage, which is much finer than that associated with ICP etching and ion implantation (13). These results indicate that SIPM is effective for mapping the process induced damages with high sensitivity.

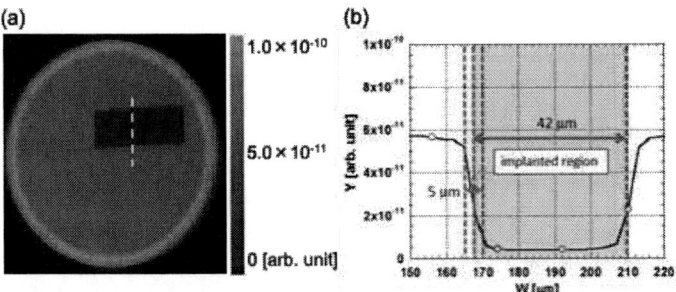

Figure 7. (a) Y image and (b) line-profile of the selective ion-implanted Ni/n-GaN Schottky contact. After Reference 10, Copyright 2016, Japan Society of Applied Physics.

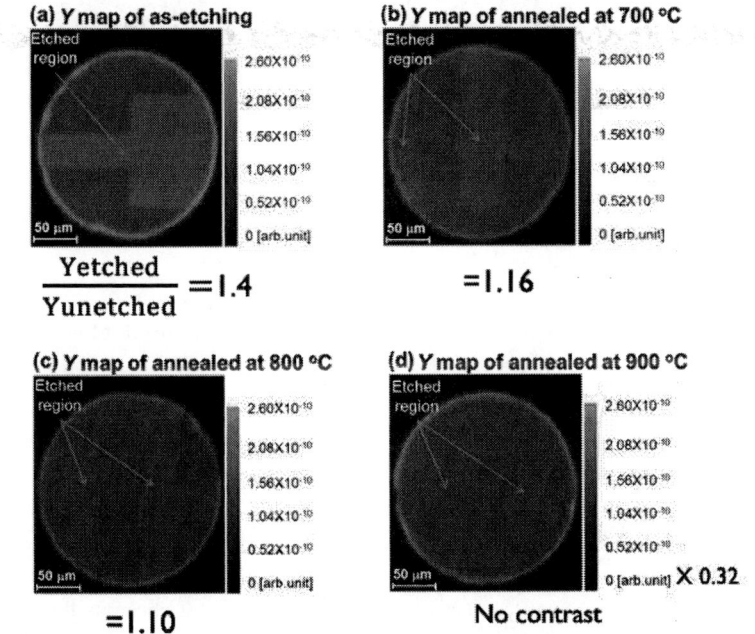

Figure 8. Y images of the selective ICP-etched Pd/n-GaN Schottky contact (a) before and after annealing at (b) 700, (c) 800, and (d) 900° C. After Reference 12, Copyright 2017, Japan Society of Applied Physics.

D: Crystal Quality and Metal Grain Boundaries

SIPM is also available for the contacts formed on a semiconductor surface with poor crystal quality. Among the SiC poly-types, 3C-SiC has the advantages of an isotropic crystal structure and high electron and hole mobilities for electron device applications. However, owing to a lack of 3C-SiC bulk crystals, heteroepitaxial growth on Si, 4H-, or 6H-SiC substrates is unavoidable. Thus, the crystal quality of 3C-SiC is not as good as those of 4H- and 6H-SiC. We measured Ni Schottky contacts on p-3C-SiC layers grown on 4H or 6H-SiC substrates (14,15). As shown in Figure 9, the sample surface consists of 3C-SiC domains with a flat top. The domain pattern was clearly visualized in a Y map. By combining Y maps measured with red and green lasers, we found that $q\phi_B$ is smaller, and a larger recombination occurs in the boundary regions than in the flat regions.

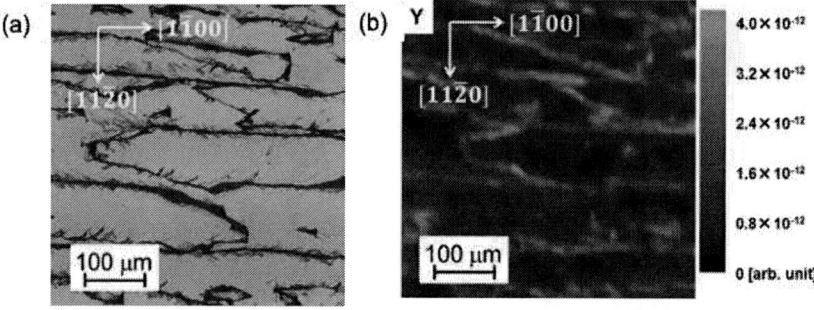

Figure 9. (a) Microscope, and Y images of a portion of the Ni contact formed on a 3C-SiC/4H-SiC layer. After Reference 14, Copyright 2017, Japan Society of Applied Physics.

Figures 10 show results of SIPM measurements for AlGaN/GaN High-Electron-Mobility Transistor （HEMT） wafers. In general, epitaxial growth of the AlGaN/GaN HEMT structure has been conducted on sapphire or SiC substrates, because these materials are chemically stable during the crystal growth. Recently, due to large wafer diameter and low cost, use of Si substrates is of interest. However, the crystal quality is not as good as HEMT wafers grown on sapphire or SiC substrates. We found typical crystal defects in on-Si wafers such as cracks and unusual crystal growth regions in the Y images in Figures 10 (b) and (c).

On the other hand, high-voltage vertical Schottky diodes have been also intensively studied because epitaxial growth of a low-carrier-density thick n-GaN drift layer became possible by using a free-standing GaN wafer. Controlling low-carrier concentration is a difficult issue due to the compensation for Si donors by C incorporation during the MOCVD crystal growth. SIPM revealed that the same pattern as the surface morphology was observed in the Y map of the Ni Schottky contact formed on the drift layer as shown in Figure 11. Since the C incorporation depended on the off-angle of the GaN surface, compensated high-resistive regions were formed. In general, characterization of electrical properties of the resistive regions is difficult, but SIPM can provide two-dimensional mapping.

As for metal deposition, vacuum evaporation and chemical vapor deposition techniques are widely used. Recently, printed electronics is known as a direct formation method for conductive patterns by conductive paste printing and sintering. The advantages of printed electronics are its milder treatment conditions, the decrease in the amount of waste of conductive ink, and the nonuse of high-vacuum equipment. We reported the basic electrical characteristics and uniformity of Ag Schottky contacts printed on n-GaN using Ag nanoink (18). The Ag electrodes were printed on the GaN substrates by drawing a single layer of Ag nanoink as shown in Figure 11 (a). The Ag nanoink was dispersed out into small particles due to a large surface tension. In the SIPM measurement, the photocurrent was detected only from a few particles which were in direct contact with the tip of a probe needle, as each particle was electrically isolated. When the Ag nanoink was overlayed four times, the particles were connected each other, and SIPM signal was detected all over the electrode as shown in Figure 11 (b).

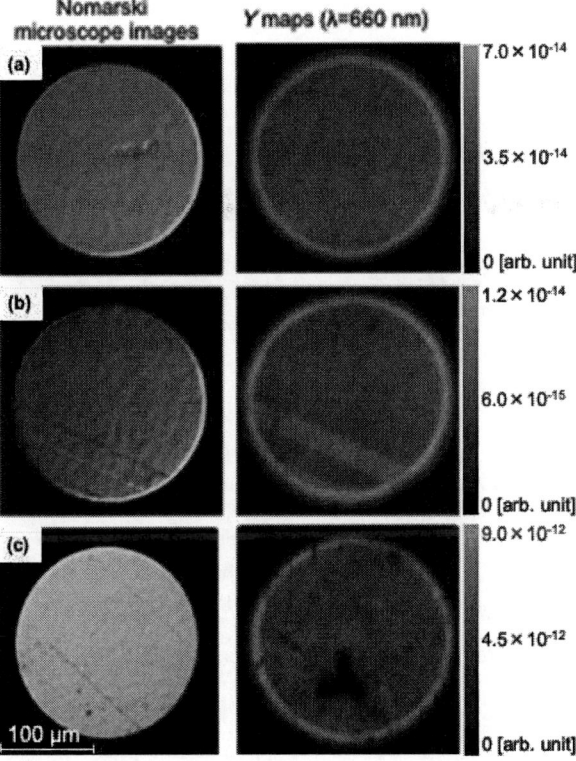

Figure 10. Microscope, and Y images of the Ni Schottky contact formed on a AlGaN/GaN HEMT structure (a) without and with (b) a crack and (c) unusual growth regions. After Reference 16, Copyright 2018, Japan Society of Applied Physics.

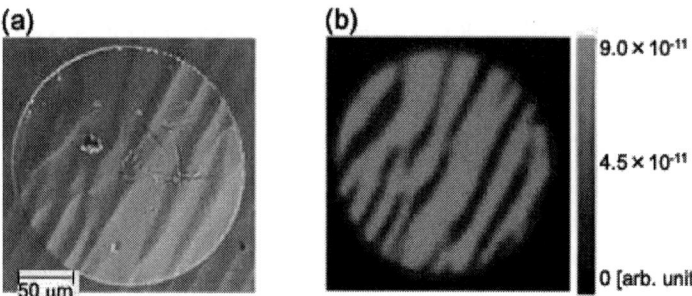

Figure 11. Microscope, and Y images of the Ni Schottky contact formed on a low-carrier-density n-GaN drift layer grown on a wavy surface of a freestanding GaN substrate. After Reference 17, Copyright 2018, John Wiley & Sons, Inc.

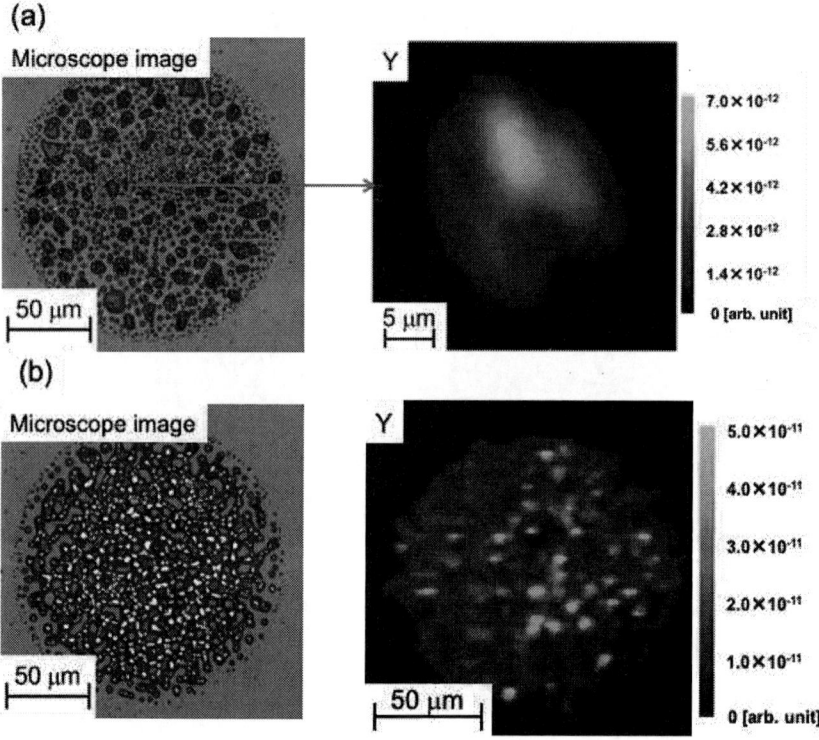

Figure 12. Microscope, and Y images of the Ag electrodes printed on the GaN substrates by drawing Ag nanoink (a) once or (b) overlaying it four times. After Reference 18, Copyright 2018, Japan Society of Applied Physics.

E: S/S and MIS Interfaces

Finally, the structural expansion of the samples is described. As long as a hetero junction and an electrical field at the interface exist, a photocurrent can be generated upon monochromatic light irradiation. We characterized p^+-Si/n^--SiC heterojunctions formed by surface-activated bonding (Figure 13) (19). In the internal photoemission spectra, a linear relationship was found between $Y^{1/2}$ and hν, and threshold energy was reasonably obtained to be 1.34 eV. In the SIPM results, Y maps were successfully obtained, and nanometer-deep scratches, which were formed on the SiC surface in wafer polishing, were sensitively visualized as a pattern.

We set an MIS structure as the next target for SIPM. We examine the breakdown by the application of high voltage to sputtered SiN$_x$ films (Figure 14) (20). After applying forward-biased voltage stress up to 30 V to Ni/ SiN$_x$ /n-SiC MIS diodes, the diode was partially degraded as shown in Figure 15. SIPM produced a clear image of the degradation pattern with a large photocurrent, which was consistent with the microscopy image. These results clarified that the partial degradation induced a leak path with a low energy barrier.

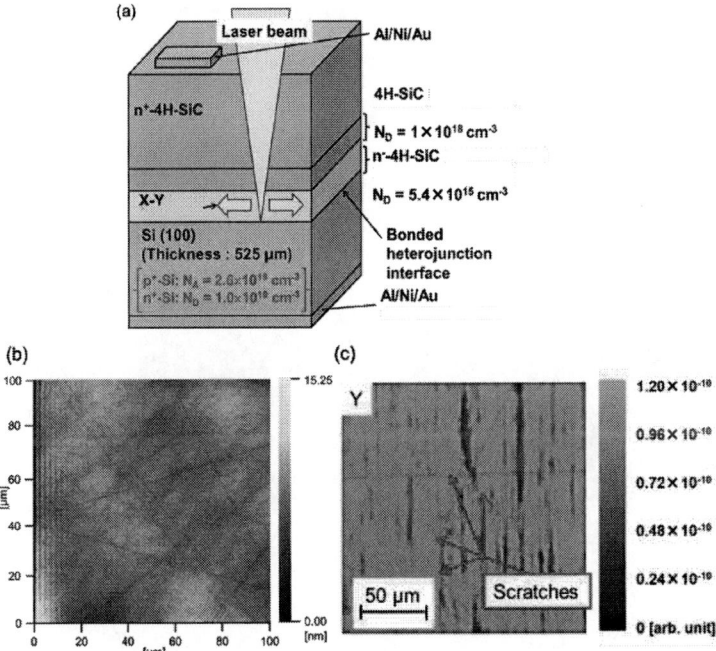

Figure 13. (a) Device structure of the bonded p^+-Si/n^--SiC heterojunctions. (b) AFM image of the SiC surface before the bonding. (c) Y image of the bonded junction. After Reference 19, Copyright 2016, Japan Society of Applied Physics.

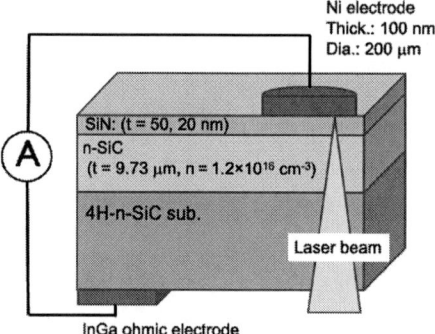

Figure 14. Device structure of the Ni/SiN$_x$/n-SiC MIS diodes.

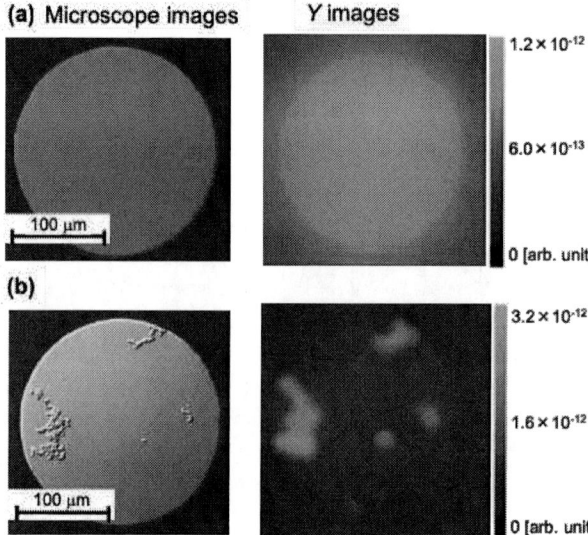

Figure 15. Microscope, and Y images of the Ni/SiN$_x$/n-SiC MIS diode (a) before and (b) after applying a reverse voltage stress of 30 V. After Reference 20, Copyright 2019, Japan Society of Applied Physics.

Conclusion

SIPM was invented about 30 years ago, and Si and GaAs Schottky diodes were preliminary targets for the measurements. As the wide-bandgap semiconductor material has been intensively studied, SIPM has also been developed with more suitable targets as shown in Fig. 16. From a 2-D characterization point of view, more microscopic observation is usually required. However, we believe that macroscopic observation, which can cover the entire electrode, is also important. Our hope is that this method will enhance not only high-power device development but also a better understanding of the basic characteristics.

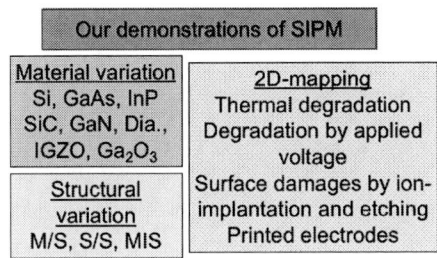

Figure 16. Perspective on our experimental results of SIPM measurements.

Acknowledgments

A part of this work was supported by a Grant-in-Aid for Scientific Research (C) 21K04135 of the Ministry of Education, Culture, Sports, Science and Technology Japan. The author would like to thank Professor Tomoyoshi Mishima and Dr. Hiroshi Ohta of Hosei University, Dr. Fumimasa Horikiri of SCIOCS Co. Ltd., Professor Masashi Kato of Nagoya Institute of Technology, Professor Naoteru Shigekawa of Osaka City University, Professor Tetsuya Suemitsu and Professor Seiji Samukawa of Tohoku University, Dr. Takashi Shinohe and Dr. Hitoshi Kanbara of FLOSFIA Inc., Professor Yasufumi Fujiwara of Osaka University, Dr. Yukiyasu Kashiwagi of Osaka Research Institute of Industrial Science and Technology, Dr. Kazushige Takechi of NLT Technologies Ltd., Dr. Kou Matsumoto of Taiyo Nippon Sanso Co., Dr. Manabu Arai of New Japan Radio Co. Ltd., Professor Mayumi B. Takeyama of Kitami Institute of Technology for fruitful discussion and providing the measurement samples. The author would also like to thank Dr. Tsugunori Okumura of President of Tokyo Metropolitan Industrial Technology Research Institute as one of the originators of SIPM.

References

1. T. Okumura and K. Shiojima, Jap. J. Appl. Phys., **28**, L1108 (1989).
2. K. Shiojima and T. Okumura, Jap. J. Appl. Phys., **30**, 2127 (1991).
3. K. Shiojima and T. Okumura, J. Crystal Growth, **103**, 234 (1990).
4. K. Shiojima and T. Okumura, Proc. of IEEE, 29th International Reliability Physics Symposium, in Las Vegas, p. 234 (1991).
5. K. Shiojima, S. Yamamoto, Y. Kihara, and T. Mishima, Appl. Phys. Exp., **8**, 046502 (2015).
6. S. Yamamoto, Y. Kihara and K. Shiojima, physica status solidi (b), **252**, 1017 (2015).
7. K. Shiojima, H. Kambara, T. Matsuda, T. Shinohe, Thin Solid Films, **685**, 17 (2019).

8. K. Shiojima and M. Shingo, physica status solidi B, **254**, 1600587 (2017).
9. K. Shiojima, M. Maeda, and T. Mishima, Jap. J. Appl. Phys., **58**, SCCD02 (2019).
10. K. Shiojima, S. Murase, S. Yamamoto, T. Mishima, and T. Nakamura, Jap. J. Appl. Phys., **55**, 04EG05 (2016).
11. S. Murase, T. Mishima, T. Nakamura, K. Shiojima, Mater. Sci. Semicon. Proc., **70**, 86 (2017).
12. A. Terano, H. Imadate, and K. Shiojima, Mater. Sci. Semicon. Proc., **70**, 92 (2017).
13. K. Shiojima, T. Suemitsu, T. Ozaki, and S. Samukawa, Jap. J. Appl. Phys., **58**, SCCD13 (2019).
14. K. Shiojima, M. Shingo, N. Ichikawa, and M. Kato, Jap. J. Appl. Phys., **56**, 04CR06 (2017).
15. K. Shiojima, N. Mishina, N. Ichikawa, and M. Kato, Jap. J. Appl. Phys., **57**, 04FR06 (2018).
16. K. Shiojima, H. Konishi, H. Imadate, Y. Yamaoka, K. Matsumoto, and T. Egawa, Jap. J. Appl. Phys., **57**, 04FG07 (2018).
17. K. Shiojima, T. Hashizume, F. Horikiri, T. Tanaka, and T. Mishima, phys. status solid B, **255**, 1700480 (2018).
18. K. Shiojima, Y. Kashiwagi, T. Shigemune, A. Koizumi, T. Kojima, M. Saitoh, T. Hasegawa, M. Chigane, and Y. Fujiwara, Jap. J. Appl. Phys., **57**, 07MA01 (2018).
19. M. Shingo, J. Liang, N. Shigekawa, M. Arai, and K. Shiojima, Jap. J. Appl. Phys., **55**, 04ER15 (2016).
20. K. Shiojima, T. Hashizume, M. Sato, and M. B. Takeyama, Jap. J. Appl. Phys., **58**, SBBC02 (2019).
21. R. H. Fowler, Phys. Rev., **38**, 45 (1931).

ECS Transactions, 104 (4) 83-91 (2021)
10.1149/10404.0083ecst ©The Electrochemical Society

**A Highly Sensitive MEMS Accelerometer Module
for Measuring Micro Muscular Sounds**

Hiroyuki Ito, Katsuyuki Machida, Noboru Ishihara, Yoshihiro Miyake, Kazuya Masu

Tokyo Institute of Technology, Kanagawa 226-8503, JAPAN

This paper presents a high-sensitivity accelerometer module to
measure micro muscular sounds. The feature of our microelectromechanical systems (MEMS) accelerometer is the use of high-density gold proof-mass structure fabricated by the multi-layer metal
technology for reducing the Brownian noise. The noise floor of the
proposed module is 10-dB better than that of a commercial one. The
measurement results showed that our module can capture small
muscular sounds which are buried in the noise of the commercial
accelerometer.

Introduction

Muscle sounds are vibrations produced by the mechanical activity of muscles, and are expected to be used in the diagnosis of diseases and fatigue (1-5). Muscle sound measurement
can be classified into two main categories: muscle displacement and muscle acceleration.
The former uses piezoelectric contact sensors, condenser microphones, or laser displacement meters, while the latter can apply capacitive accelerometers. The acceleration is the
second-order derivative of the displacement $\sin \omega t$ and has the coefficient of frequency ω^2.
In other words, when measuring high-frequency vibration such as muscle sound, it is more
advantageous to acquire acceleration because the signal strength, i.e. the signal-to-noise
ratio, can be higher. However, since the sensitivity of commercially available small microelectromechanical systems (MEMS) accelerometers is generally low, they do not offer the
above-mentioned advantage although they are compact in size. Therefore, conventional
muscle sound measurement systems are either highly-sensitive and bulky, or low-sensitiv-
ity and compact. The realization of a highly-sensitive and compact muscle-sound measurement system has a potential for contributing to the development of continuous monitoring
and early diagnosis of diseases and fatigue.

We have realized a MEMS accelerometer with much lower Brownian noise B_N than
conventional silicon-based accelerometers by using high-density gold proof-mass structure
(6,7) fabricated by the multi-layer metal technology (8). Our idea is to employ the ultra-
sensitive accelerometer for micro muscle-sound measurement. This paper presents the possibility of weak muscle-sound measurement using ultra-high-sensitivity accelerometers
based on the contents of Ref. (9).

Module Development

Noise Analysis of the Module

In order to improve the sensitivity of accelerometers, it is necessary to reduce the total noise (T_N) while increasing the gain over the signal so that small accelerations can be detected. The overall noise (T_N) of the accelerometer module consists of the Brownian noise (B_N) (10), which is the mechanical noise of the MEMS device, and the electrical noise (E_N) of the interface circuit, and is given by (11)

$$T_N = \sqrt{B_N{}^2 + E_N{}^2} = \sqrt{B_N{}^2 + \left(\frac{\Delta C_{min}}{S_m}\right)^2} \qquad [1]$$

where ΔC_{min} is the minimum capacitance change that can be detected by the interface circuit, and S_m is the capacitance change over input acceleration of the MEMS accelerometer device. Our prototype utilizes a commercially available capacitance-digital converter (AD7745, Analog Devices Inc.) as the interface circuit, and its capacitance sensitivity ΔC_{min} is 4.3 aF/Hz$^{1/2}$ at the output data rate of 90.9 Hz according to the spec sheet (12). When the target of total noise T_N is set to be below 100 μG/Hz$^{1/2}$, which is smaller than those of conventional sensors (13), the relationship between Brownian noise B_N and acceleration-capacitance conversion gain S_m as the target value of the MEMS device can be shown as in Fig. 1.

Design and Evaluation of MEMS Accelerometer Device

Figure 2 shows a cross-sectional view of the single-axis high-sensitivity MEMS accelerometer device with a gold proof-mass fabricated using multi-layer metal technology (8). It consists of a proof mass, a fixed electrode on a Si substrate, springs connecting the proof mass and the Si substrate, and a stopper to limit the moving range of the proof mass. The change in the proof mass position caused by acceleration is detected by the change in capacitance between the proof mass and the fixed electrode. This device can detect

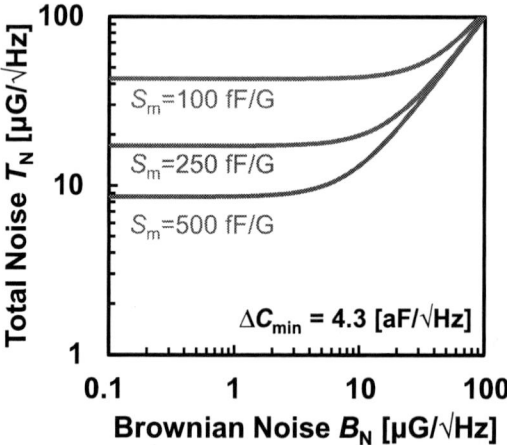

Figure 1. Relationship among total noise, Brownian noise B_N and acceleration-capacitance conversion gain S_m (9).

acceleration in the direction perpendicular to the plane of the Si substrate. The device was designed to have a Brownian noise B_N of less than 200 nG/Hz$^{1/2}$ and acceleration-capacitance conversion gain S_m of more than 250 fF/G to provide enough margin from the target value of total noise which is less than 100 μG/Hz$^{1/2}$. In addition, the resonant frequency of the MEMS device was designed to be sufficiently higher than the bandwidth of the interface circuit, which is about 50 Hz. Figure 3 shows a micrograph of the prototype device, which has a chip area of 4 mm square and is packaged in a module.

Ring-down tests were conducted on this device with the measurement system shown in Fig. 4. A pulse with an amplitude of 4 V is applied between the proof mass and the fixed

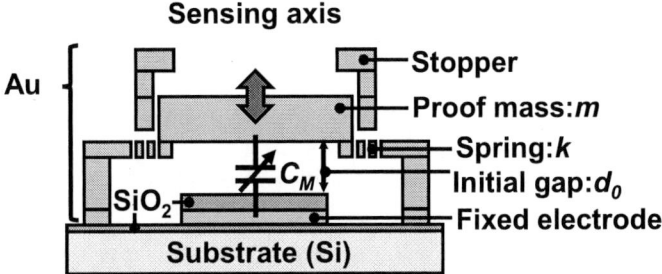

Figure 2. Cross section of the single-axis MEMS accelerometer with a gold proof mass (9).

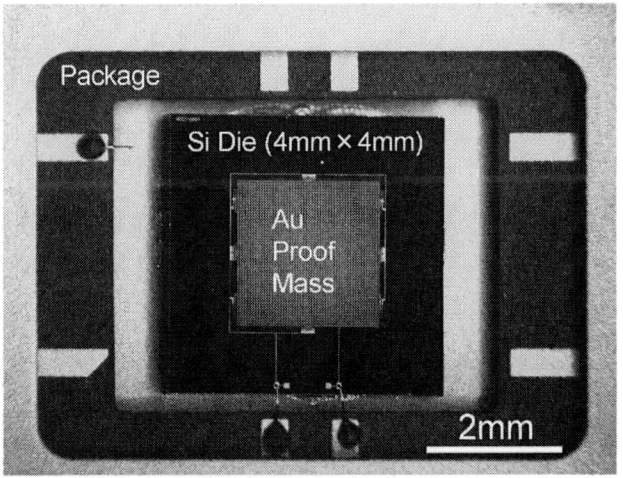

Figure 3. Micrograph of the developed MEMS accelerometer (9).

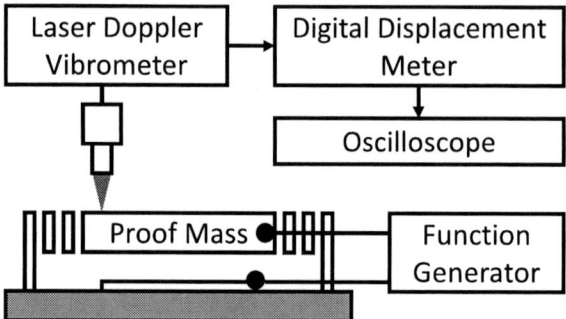

Figure 4. Measurement system for ring-down tests.

electrode by a function generator (33511B, Keysight Technology), and the proof mass is momentarily brought closer to the fixed electrode side by electrostatic force, and then vibrates freely. To avoid pull-in, the pulse width is made sufficiently smaller than the reciprocal of the resonance frequency. The free vibration of the proof mass is measured by a laser Doppler vibrometer (LV-1800, Ono Sokki Co., Ltd), and Fig. 5 shows the time-domain waveform of it. Resonant angular frequency ω_{res} and quality factor Q can be found by fitting the time-domain waveform of the proof-mass displacement x shown in the following equation (14):

$$x = e^{-\zeta \omega_{res} t} \left\{ x_0 \cos(\omega_d t) + \frac{v_0 + \zeta \omega_{res} x_0}{\omega_d} \sin(\omega_d t) \right\}$$ [2]

where ζ is the damping factor of the vibration, and the quality factor can be obtained from $Q = 1/\zeta$. x_0 is the initial position of the proof mass immediately after the pulse is applied. v_0 is the initial velocity of the proof mass, and $\omega_d = \omega_{res}\sqrt{1 - \zeta^2}$. Then, the resonant frequency f_{res} and quality factor Q of the fabricated MEMS device were found to be 380 Hz and 13.2, respectively. From these values and the Brownian noise equation (15)

$$B_N = \frac{1}{9.8} \sqrt{\frac{4 k_b T \omega_{res}}{mQ}},$$ [3]

the Brownian noise B_N of the MEMS device can be evaluated to be 198 nG/Hz$^{1/2}$, where k_b is Boltzmann's constant (1.38×10^{-23} J/K), T [K] is room temperature. m is the mass of the proof mass, and the design value 8.43×10^{-7} g was used.

The MEMS device was excited with a sinusoidal acceleration of 19.9 Hz by a vibration excitation system (WaveMaker05, Asahi Seisakusho Co.), and the change in capacitance at the frequency of 19.9 Hz was measured by a semiconductor parameter analyzer (B1500A,

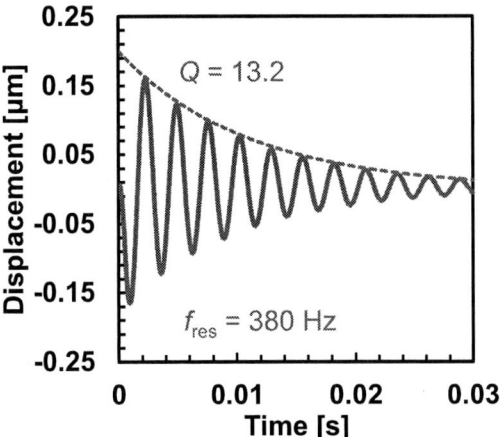

Figure 5. Ring down measurement result of the MEMS device (9).

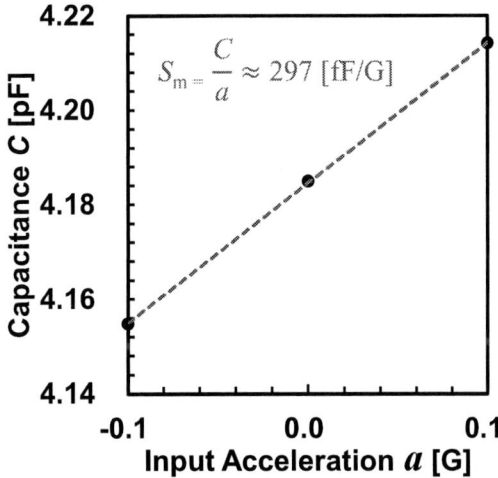

Figure 6. Measured capacitance a function of input acceleration (9).

Keysight Technology). As shown in Fig. 6, the capacitance changes linearly with acceleration input from -0.1 G to 0.1 G, and the acceleration-capacitance conversion gain S_m of

297 fF/G can be estimated from this slope. Thus, it was confirmed that measured B_N and S_m meet the target value.

Ultra-Sensitive Accelerometer Module for Muscle-Sound Measurement

Figure 7 shows a block diagram of the accelerometer module prototype for muscle sound measurement. Our ultra-sensitive MEMS accelerometer consists of the accelerometer device shown in Fig. 3 and a commercially available capacitance-digital converter (AD7745, Analog Devices Inc.). To verify the benefits of the high-sensitivity accelerometer, a commercial inertial sensor MPU6050 (13) is implemented on the same printed circuit board. Each measurement data is sent to the microcontroller through the I2C interface, and then to the personal computer through USB (Universal Serial Bus) communication.

Figure 8 is a photo of the prototype, and the size of the module is 15 mm × 20 mm. Our MEMS device and the commercial accelerometer are mounted on the front side, and the capacitance-digital converter is implemented on the back side to minimize the wiring length between MEMS device and to reduce the module size. The module is packaged in a small and lightweight plastic case and attached to the finger with a rubber band. The microcontroller and a battery for the sensors are worn on the arm, and I2C communication

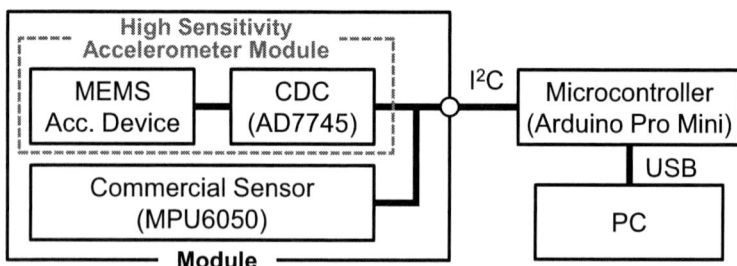

Figure 7. Block diagram of the prototype system for muscle-sound measurement (9).

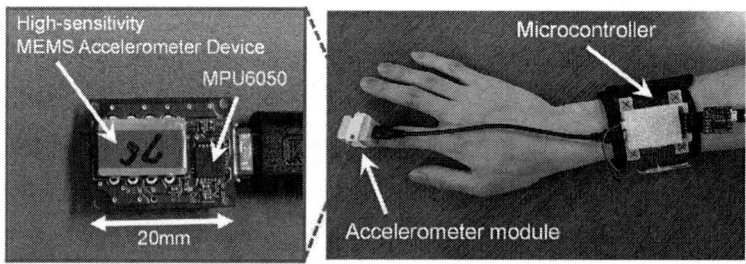

Figure 8. Photo of the prototype system for muscle sound measurement (9).

and power supply are provided through cables. In order to reduce the influence of noise from the power source, the prototype uses a battery to supply power for the sensors.

Experimental Results

In order to evaluate the noise floor, the module and the microcontroller were placed on a static floor in a horizontal position. Figure 9 shows the output acceleration spectral density of the commercial product and our high-sensitivity accelerometer. The measurement time is 60 seconds, and the vertical axis is the ratio to 1 G, i.e. 0 dB = 1 G. The noise floor of our accelerometer is less than 100 $\mu G/Hz^{1/2}$, while the noise floor of commercial accelerometers is about 400 $\mu G/Hz^{1/2}$. This means that our sensor meets the design target of less than 100 $\mu G/Hz^{1/2}$ and has a noise floor at least 10 dB better than that of commercial one.

The module was worn on a forefinger as shown in Fig. 8. The subjects were three healthy young males, and one of their data is shown in Fig. 10. In this measurement, the forearms and hands of the subjects were placed on a table. The measurement time was 60 seconds, and the output acceleration spectral density was calculated by excluding the first 10 seconds of data. The signal around 10 Hz is considered to be a physiological tremor observed even in healthy people at any condition (16). Our high-sensitivity accelerometer has succeeded in capturing weak vibrations in the frequency range of 20Hz to 40Hz, which is in the muscle sound frequency band (5), whereas the conventional MEMS accelerometers cannot measure it due to the inherent noise floor (9).

Conclusion

This paper presented the possibility of measuring weak muscle noise using an ultra-sensitive accelerometer. By using a high-density gold proof-mass structure with multi-layer metal technology, the Brownian noise of the single-axis accelerometer was reduced, and a noise floor 10 dB lower than that of commercial accelerometers was achieved. By wearing our accelerometer on the forefinger, weak muscle sound in the 20 Hz - 40 Hz band, which is buried in noise by conventional accelerometer, could be detected.

Acknowledgments

The authors would like to thank S. Iida and T. Konishi with NTT-AT Corp. for technical discussions, and T. Koga and T. Ichikawa for design and measurement supports. This work was supported by JST CREST Grant Number JPMJCR1433, Japan, JSPS KAKENHI Grant Number 19K05232, and JST A-STEP Grant Number JPMJTM20N3.

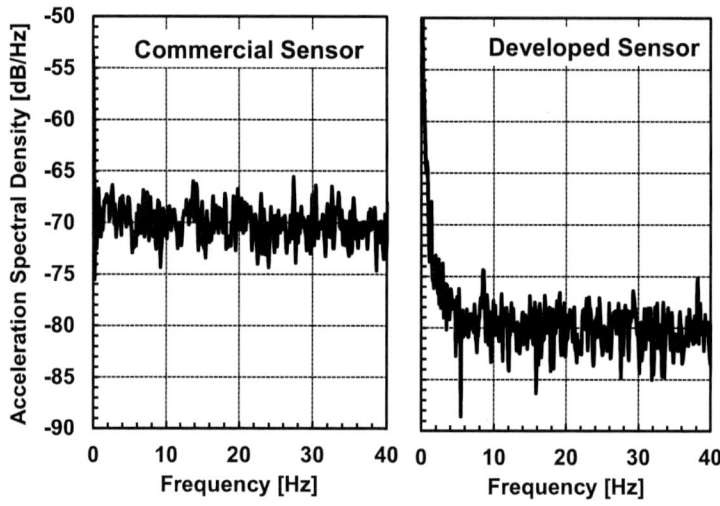

Figure 9. Measured acceleration spectral density on a static floor (9).

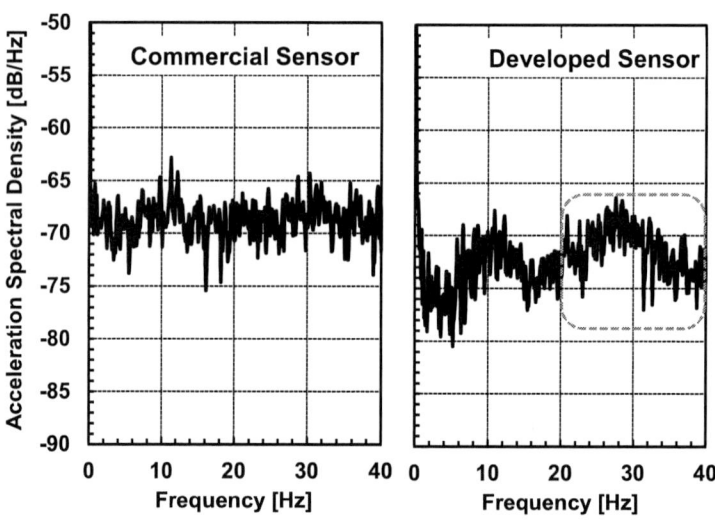

Figure 10. Measured acceleration spectral density on a forefinger (9).

References

1. F. Esposito, D. Malgrati, A. Veicsteinas, and C. Orizio, *Eur. J. Appl. Physiol.*, **73**, (1996).
2. P. Brown, *Lancet*, **349**, (1997).
3. C. Orizio, M. Gobbo, B. Diemont, F. Esposito, and A. Veiscteinas, *Eur. J. Appl. Physiol.*, **90**, (2003).
4. J. Marusiak, A. Jaskolska, K. Kisiel-Sajewicz, G. H. Yue, and A. Jaskolski, *J. Electromyogr. Kinesiol.*, **19**, (2009).
5. M. Ouamer, M. Boiteux, M. Petitjean, L. Travens, and A. Sales, *J. Biomech.*, **32**, (1999).
6. D. Yamane, T. Konishi, T. Matsushima, K. Machida, H. Toshiyoshi, and K. Masu, *Appl. Phys. Lett.*, **104**, (2014).
7. D. Yamane, T. Konishi, T. Matsushima, H. Toshiyoshi, K. Masu, and K. Machida, *J. Appl. Phys.*, **54**, (2015).
8. K. Machida, S. Shigematsu, H. Morimura, Y. Tanabe, N. Sato, N. Shimoyama, T. Kumazaki, K. Kudou, M. Yano, and H. Kyuragi, *IEEE Trans. Electron Devices*, **48**, (2001).
9. T. Koga, T. Ichikawa, N. Tanaka, T. Ogata, H. Ora, D. Yamane, N. Ishihara, H. Ito, M. Sone, K. Machida, Y. Miyake, and K. Masu, *in Proc. IEEE Biomedical Circuits and Systems Conf.*, (2019).
10. M. Lemkin and B. E. Boser, *IEEE J. Solid-State Circuits*, **34**, (1999).
11. B. V. Amini and F. Ayazi, *J. Micromech. Microeng.*, **15**, (2005).
12. AD7745/AD7746: 24-Bit Capacitance to Digital Converter with Temperature Sensor Data Sheet (Rev. 0)
13. MPU-6000 and MPU-6050 Product Specification (Rev. 3.4)
14. H. Hosaka, Introduction to Mechanical Vibrations (in Japanese), University of Tokyo Press, (2005)
15. T. B. Gabrielson, *IEEE Trans. Electron Devices*, **40**, (1993).
16. R. Herbert, *J. Physiol.*, **590**, (2012)

92

**Nanoscale Probing of Field-Driven Ion Migration in TaO$_x$
for Neuromorphic Memristor Applications**

A. Tsururmki-Fukuchi[a], T. Katase[b], H. Ohta[c], M. Arita[a], and Y. Takahashi[a]

[a] Faculty of Information Science and Technology, Hokkaido University,
Sapporo 060-0814, Japan
[b] Laboratory for Materials and Structures, Institute of Innovative Research,
Tokyo Institute of Technology, Yokohama 226-8503, Japan
[c] Research Institute for Electronic Science, Hokkaido University,
Sapporo 001-0020, Japan

> Stable operations as resistive switching memory were
> demonstrated in amorphous TaO$_x$ (a-TaO$_x$) thin films with very
> flat surfaces by conductive atomic force microscopy (c-AFM). The
> a-TaO$_x$ thin films fabricated on glass and Nb-doped SrTiO$_3$
> substrates by pulsed laser deposition showed high surface flatness
> with root-mean-square roughness of less than 0.2 nm. The c-AFM
> observations on the surfaces revealed that the resistive switching in
> a-TaO$_x$ causes almost no change in the topographic structures, and
> the significant structural deformation appears after the electrical
> breakdown by longer-range migration of the constituent ions. The
> possible mechanisms of the resistive switching phenomena were
> discussed based on the changes in the topographic structures and
> conduction states.

Introduction

In recent years, increasing attention has been paid to the physical mechanisms of the
field-driven ion migration in amorphous metal-oxide semiconductors as it has been
considered an important principle of resistive random access memory (ReRAM), which is
known as a promising class of emerging memory device. In the basic mechanisms of
ReRAMs, reversible and fast changes in the electrical resistance are caused by the field-
driven ion migration and resultant changes of the valence states of the switching oxide,
and their promising functions as nonvolatile or neuromorphic (brain-inspired) memory
devices are brought through the resistance changes. Among the amorphous metal oxides
with resistive-switching functionality, amorphous TaO$_x$ (a-TaO$_x$) has been considered a
material of particular importance because it can offer very high-performance nonvolatile
memory characteristics to ReRAM devices (1, 2) and has been commercialized as the
practical material (3).

To improve the process integration and further expand the application, microscopic
investigations have been extensively conducted for the ion migration in a-TaO$_x$ in the last
decade. Conductive scanning probe microscopy has been considered a suitable way for
the investigation as this method enables both electrical measurements of resistive
switching phenomena and nanoscale characterizations of the ion migrations (4–14). In a-
TaO$_x$, structural changes caused by the resistive switching to the low resistance state

(called "set" operation) has been observed at the nanometer scale, and the involved mechanisms of the ion migration has been discussed (15, 16). These previous works have also showed that the probe microscopy observations of resistive switching phenomena are strongly influenced by the surface morphologies of the investigated oxides. It has been demonstrated that the current–voltage (I–V) characteristics, spatial distributions of the conductance, and structural changes by ion migration are strongly dependent on the structural distributions of initially included grains and defects in the thin films of FeO_x, NiO_x, $SrTiO_x$, HfO_x, and WO_x (4–6, 8, 9, 14). However, these results suggests that the large contributions from the surface morphologies may hinder the processes of the ion migration which are actually responsible for the resistive switching. In other words, if the analysis can be conducted for a-TaO_x thin films with high flatness and uniform distributions of the ionic defects, important clues for the unexplained mechanisms of the resistive switching may be obtained. In this study, therefore, we fabricated very flat thin films of a-TaO_x by pulsed laser deposition (PLD) method and investigated the ion migration in the resistive switching phenomena by conductive atomic force microscopy (c-AFM) in order to understand the detailed underlying mechanisms of TaO_x-based ReRAMs.

Fabrication of a-TaO_x Thin Films with Very Flat Surfaces

Previous studies reported that the surface morphology of an a-TaO_x thin film deposited by sputtering at room temperature is generally rough, and spatial variations of the ionic concentrations are included in the film (17). The rough surface morphologies of a-TaO_x thin films have been attributed to the limited surface migration of the adatoms under the room-temperature deposition and self-shadowing effect of the aggregated atoms based on the structure zone model (18), which is a well-accepted model for low-temperature film deposition. At the same time, this model tells us that highly flat thin films of a-TaO_x, which are suitable for the c-AFM analysis, can be fabricated by increasing the kinetic energies of the incident Ta atoms during the deposition. Based on this assumption, we conducted the deposition of a-TaO_x thin films by the PLD method in this study, by which large kinetic energies of a few hundred eV can be provided for the incident atoms (19).

Figure 1(a,b) shows the surface morphologies of a bare glass substrate and an a-TaO_x(50 nm) thin film deposited on the glass substrate. The deposition of the a-TaO_x films was conducted at room temperature by the PLD method using Nd:YAG laser (with a wavelength of 532 nm) at a laser power of 15 mJ, and the oxygen partial pressures (P_{O2}) during the deposition were in a range of 5×10^{-5}–10 Pa. The thicknesses of the a-TaO_x thin films were determined from the X-ray reflectivity. As shown in the AFM images, the a-TaO_x thin film deposited by the PLD at $P_{O2} = 1.5$ Pa showed a very flat surface morphology and the root mean square roughness (R_{rms}) was decreased from that of the used glass substrate. According to the structure zone model (18), the surface roughness of a deposited thin film is expected to increase with increasing the gas pressure during the deposition because of the reduced surface mobility of incident atoms. However, we observed that the a-TaO_x thin films keep the high surface flatness to a relatively high P_{O2} of 10 Pa owing to the high kinetic energies under the PLD. The R_{rms}s of the deposited a-TaO_x thin films were comparable with or smaller than that of the glass substrates in the P_{O2} range of 5×10^{-5}–10 Pa (Fig. 1(c)).

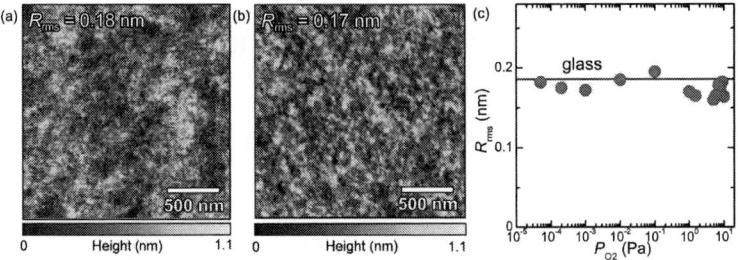

Figure 1. AFM topographic images of (a) a glass substrate and (b) an a-TaO$_x$(50 nm) thin film deposited under P_{O2} = 1.5 Pa on the glass substrate. (c) P_{O2} dependence of the R_{rms} for the a-TaO$_x$ (50 nm) thin films. Horizontal line in the figure represents the R_{rms} value of the glass substrate.

For the c-AFM analysis of the resistive switching, we used Nb-doped SrTiO$_3$ (Nb:STO) (001) conductive substrates with a Nb-doping concentration of 1.0%, on which high surface flatness is available by chemical mechanical polishing and a low-resistance ohmic contact can form with an a-TaO$_x$ film through the interfacial redox reactions (20). We observed that the PLD-deposited a-TaO$_x$/Nb-STO (001) thin films also have very flat surface morphologies with small R_{rms}s of <0.2 nm. Similarly to the a-TaO$_x$/glass thin films, the R_{rms} values of the a-TaO$_x$/Nb-STO (001) thin films were comparable with or decreased from that of the used Nb:STO (001) substrate. As shown in Fig. 2(a,b), R_{rms}s of 0.15 nm and 0.16 nm were measured for an a-TaO$_x$(8.6 nm) thin film and Nb:STO (001) substrate, respectively. We observed that the very small R_{rms} of <0.2 nm, which is at an atomic level, are not significantly influenced by the increase of the film thickness of a-TaO$_x$ up to over 100 nm.

The measurement setup used for the c-AFM analyses is shown as Fig. 2(c). In electrical measurements of resistive switching phenomena by scanning probe microscopy, previous works showed that the large current flow during the resistive switching often causes melting and blunting of the conductive tip (10, 11), which may cause undesirable chemical and mechanical interactions between the tip and film materials (measurement contamination). To prevent such tip degradation, we used Rh-coated conductive tips which have a high melting point of 2237 K and chemical/mechanical stability under the measurements. We confirmed that the electrical measurements in this study caused no detectable chemical and mechanical contamination between the Rh-coated tips and a-TaO$_x$ thin films (such as electrochemical reactions, alloying, and electrical deposition of the metal elements), based on their Auger electron spectroscopy analyses (data not shown). Additionally, to close the large gap in the orders of magnitude of the typical measurement currents between c-AFM experiments (10^{-8} A) and practical ReRAM devices (10^{-5} A) (3), the electrical measurements for the resistive switching were conducted via an external source measure unit (Keysight B1500A), and using the unit, compliance currents (I_c) were set for the a-TaO$_x$ thin films during the I–V measurements.

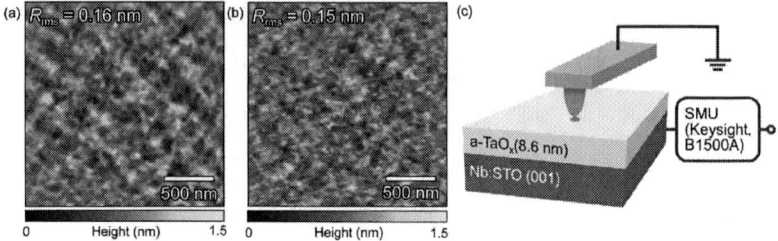

Figure 2. AFM topographic images of an (a) Nb:STO (001) substrate and (b) a-TaO$_x$(8.6 nm) thin film deposited on the Nb:STO substrate by PLD. (c) Schematic illustration of the measurement setup for c-AFM analysis for the very flat thin films of a-TaO$_x$/Nb:STO (001). The measurements were conducted in air.

Resistive Switching Operations of a-TaO$_x$/Nb:STO by c-AFM

By applying positive voltages of >3.0 V to the Nb:STO substrate under I_c of >1.0 μA, resistive switching behaviors to the low resistance state were observed in the a-TaO$_x$ thin films (Fig. 3(a)), which correspond to "set" operations in ReRAM devices. Interestingly, no apparent change was detected in the topographic and current AFM images of the a-TaO$_x$ surfaces after the "set" operations (Fig. 1(b)), while resistive switching with a large resistance ratio of >10^4 was induced in the I–V measurements. When additional positive voltages were applied to the a-TaO$_x$ thin film after the set switching, the resistance of the a-TaO$_x$ film was further decreased (Fig. 3(c)), and no further resistance switching was caused in the a-TaO$_x$ thin film after the second voltage sweeping. Based on the loss of the resistive switching function, we can consider the resistance decrease by additional voltage applications as the breakdown operations in ReRAMs which are generally caused by excessive electrical inputs (called "hard breakdown") (21). After the breakdown operations, we observed the formation of a very characteristic structure of a two-stage mound on the a-TaO$_x$ thin film (Fig. 3(d)). The current image of the a-TaO$_x$ thin film indicated that a sharp conductive spot with a diameter of 30 nm appeared at the center of the structural deformation (Fig. 3(d)). The conduction spot with a dimension of a few 10 nm is similar to that previously observed for other resistive switching oxides by c-AFM (6, 8, 9). However, our results suggests that the observed spot is the origin of the breakdown operation of a-TaO$_x$ rather than the stable resistive switching, and the actual conduction path involved in the resistive switching may have a smaller dimension or be transient.

In the a-TaO$_x$ thin films, the resistance ratio in the resistive switching showed a dependence on the I_c during the I–V measurements. In the I–V characteristics measured at I_c = 1.0 μA (Fig. 3(e)), the a-TaO$_x$ thin film showed a smaller resistance ratio of around 10^2. This suggests that the smaller amount of ion migration was caused by the voltage sweeping at I_c = 1.0 μA, and an analog resistive switching function (22), which is a function of critical importance in the neuromorphic applications, may be available in the a-TaO$_x$ thin films by controlling the I_c. When additional voltages were applied for the a-TaO$_x$ thin film after the incomplete set operation, resistive switching was again observed, but the switching voltage was increased from the initial measurement (Fig. 3(e)). After

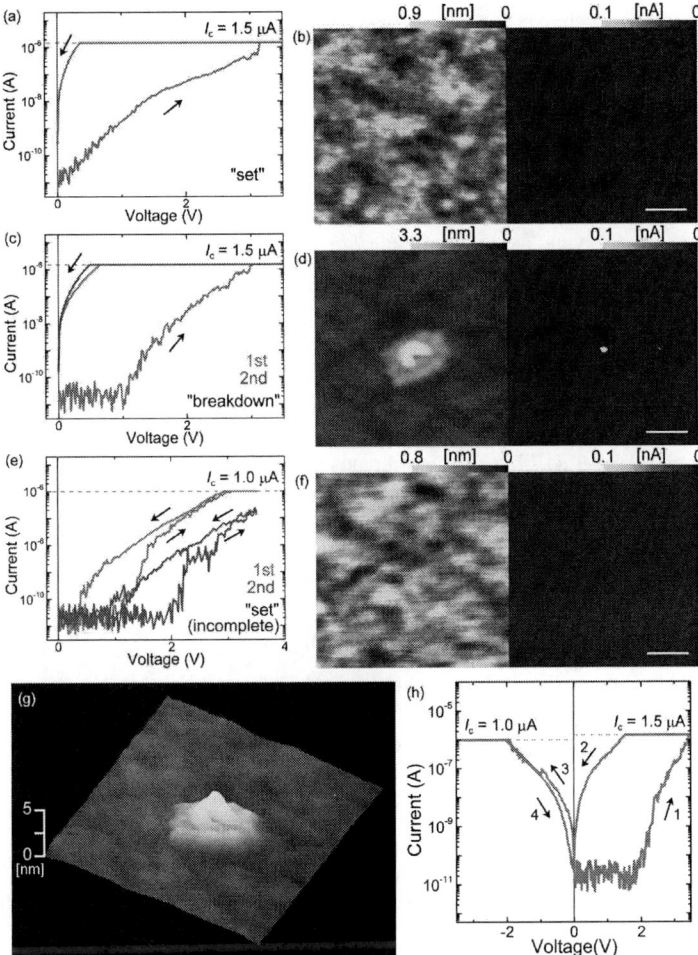

Figure 3. (a) I–V characteristics measured by c-AFM with I_c = 1.5 μA (single sweeping) and the (b) (left) topographic and (right) current images observed after the "set" operation for an a-TaO$_x$(8.6 nm) thin film. (c) I–V characteristics measured by c-AFM with I_c = 1.5 μA (double sweeping) and the (d) (left) topographic and (right) current images observed after the "breakdown" operation for an a-TaO$_x$(8.6 nm) thin film. (e) I–V characteristics measured by c-AFM with I_c = 1.0 μA (double sweeping) and the (f) (left) topographic and (right) current images observed after the "incomplete set" operation for an a-TaO$_x$(8.6 nm) thin film. The current images were measured at a reading voltage of +1.0 V. The scale bars in the AFM images are 200 nm. (g) Three-dimensional plot of the topographic image of the a-TaO$_x$ thin film after the breakdown operation. The scan size is 900 nm × 900 nm. (h) I–V characteristics measured by c-AFM in a voltage-sweeping sequence of 0 V → +3.5 V → −3.5 V → 0 V ("reset" operation) for an a-TaO$_x$(8.6 nm) thin film.

the twice incomplete set operations, no significant changes in the structures and conduction states were observed in the AFM images (Fig. 3(f)) as well as after the single complete set (Fig. 3(b)). Therefore, our results showed that the structural changes of a-TaO_x became observable for the first time after the breakdown operation even on the very flat surfaces, while the dimensions of the deformation (2.5 nm in the out-of-plane direction and 250 nm in the in-plane direction) were large after the breakdown and comparable to the film thickness (Fig. 3(g)). In contrast, the switching to the high resistance state (called "reset" operation in ReRAM devices) was observed after the set operations by applying negative voltages to the films (Fig. 3(h)), while no notable structural changes were induced in the a-TaO_x thin films by the first set operation.

The large change in the resistance observed in the set operation (Fig. 3(a)) indicates that significant amount of ion migration was caused by the voltage sweeping although little structural deformation was observed on the surface (Fig. 3(b)). The characteristic two-plateau structure formed by the breakdown operation (twice sweeping of positive voltages) (Fig. 3(d,g)), which is unlikely to be formable by single voltage stress, also suggests that the a-TaO_x thin film has already experienced nonnegligible amount of ion migration in the first set operation. Based on these considerations, the structural insensitivity of the a-TaO_x thin films in the resistive switching can be explained from the high ionic conductivity and large difference in the ionic radius of the constituent ions. As an amorphous metal oxide, a-TaO_x has been suggested to have a high ionic conductivity at room temperature both for Ta^{5+} (15, 23) and O^{2-} (24) in previous observations. In addition, when the ionic radii of Ta^{5+} and O^{2-} are assumed to be 0.064 nm and 0.140 nm (25), respectively, the volume occupied by the Ta^{5+} ions in a-TaO_x is expected to be about 10% as compared with that by the O^{2-} ions. This indicates that the volume changes caused by the migration of Ta^{5+} ions can be readily compensated by the migration of 10% amount of O^{2-} ions in a-TaO_x, while the charge states are not fully compensated.

Under these circumstances, the ion migration involved in the observed resistive switching are explained as follows. In the set switching by the application of positive voltages (Fig. 3(a)), Ta^{5+} and O^{2-} ions will be moved by field-driven drift at the initial stage, and the concentrations of Ta ions will be gradually enriched from the tip/film interface by the drift-driven mass flux (J_d) in the out-of-plane direction (Fig. 4(a)). The volume increase by the Ta accumulation can be canceled out by the drift of O^{2-} ions in the reverse direction, and O^{2-}-enriched a-TaO_x will be formed in the surrounding region. The set switching will be made when a conduction path penetrating the film is completed by the Ta-enriched a-TaO_x with reduced resistivity without causing significant deformation in the film structure. The structural invariability of a-TaO_x in the set switching is consistent with the recent observation by scanning transmission electron microscopy (23). At this stage, the conduction path is supposed to have a lateral dimension less than 10 nm, which is not observed in the current mapping by c-AFM, or compositionally unstable to such an extent as to be reoxidized in the c-AFM scanning. A similar volatility of the conduction path has recently been reported in a-HfO_x thin films after the resistive switching (12). When additional voltages are applied to the a-TaO_x thin film in the low resistance state (Fig. 3(c)), the film is subjected to increased current stress, and increased mass flux (J_S) will be caused by Soret effect (26) in the in-plane direction. By the further concentration of Ta ions into the conduction path, the path will start to deform in the out-of-plane direction because of the increased migration of Ta ions (15, 16). After that, the film will turn into the breakdown state with accompanying further

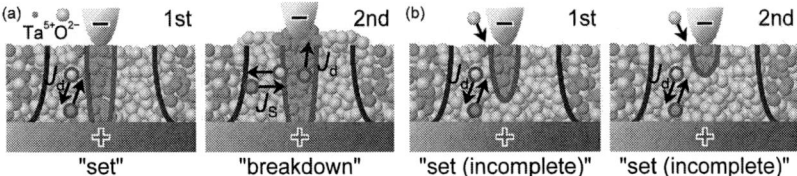

Figure 4. Schematic illustrations of the ion migration processes in a-TaO$_x$ thin films for the (a) set and breakdown and (b) incomplete set operations by c-AFM. The actual size ratio between Ta^{5+} and O^{2-} ions is shown at upper left of (a).

O^{2-}enrichment in the surrounding region (right panel of Fig. 4(a)). The increased J_s will not be generated when the conduction path is not completed at a smaller I_c (Fig. 4(b)). Therefore, in the incomplete set operation, the resistance and switching voltages may be increased in the second voltage sweeping (right panel of Fig. 4(b)) due to the O^{2-} concentrations increased by electrochemical oxidation from air.

Conclusions

Ion migration analyses involved in the resistive switching were conducted for highly flat thin films of a-TaO$_x$/Nb:STO (001) by c-AFM. Unlike other deposition method, the room-temperature-deposited films of a-TaO$_x$ by the PLD method showed high structural uniformity and a flat surface morphology with R_{rms} < 0.2 nm owing to the high deposition energies. In the I–V characteristics by c-AFM, clear, reproducible resistive switching phenomena were observed for the a-TaO$_x$ thin films. The structural and conductance mapping by c-AFM showed that the resistive switching to the low resistance state (set operation) causes almost no structural deformation for a-TaO$_x$, and the conduction path which contributes to the resistive switching has an unobservable size in the scanning measurements (<10 nm in the lateral direction) or chemically unstable after the set switching. After an a-TaO$_x$ thin film reached the breakdown state, characteristic structural deformation and a 10 nm-sized conduction path were observed on the surface, which suggest the occurrence of longer-range ion migration by thermal diffusion. These results suggested that prevention of long-range ionic migration is crucial for the stable operations of a-TaO$_x$-based ReRAMs, and to observe the actual conduction path in the resistive switching and understand the nature, a-TaO$_x$ thin films with further improved flatness (e.g., by substrate surface treatment) will be required.

Acknowledgments

This work was financially supported by the Japan Society for the Promotion of Science (JSPS, KAKENHI No. 19K04484) organized by the Ministry of Education, Culture, Sports, Science, and Technology (MEXT), Japan. Part of this work was conducted under the Nanotechnology Platform by MEXT and the Collaborative Research Project of Laboratory for Materials and Structures, Institute of Innovative Research, Tokyo Institute of Technology.

References

1. Z. Wei, Y. Kanazawa, K. Arita, Y. Katoh, K. Kawai, S. Muraoka, S. Mitani, S. Fujii, K. Katayama, M. Iijima, T. Mikawa, T. Ninomiya, R. Miyanaga, Y. Kawashima, K. Tsuji, A. Himeno, T. Okada, R. Azuma, K. Shimakawa, H. Sugaya, T. Takagi, R. Yasuhara, K. Horiba, H. Kumigashira, and M. Oshima, *Tech. Dig. IEEE Int. Electron Dev. Meet.*, 293–296 (2008).
2. M.-J. Lee, C. B. Lee, D. Lee, S. R. Lee, M. Chang, J. H. Hur, Y.-B. Kim, C.-J. Kim, D. H. Seo, S. Seo, U-I. Chung, I.-K. Yoo, and K. Kim, *Nat. Mater*, **10**, 625–630 (2011).
3. Y. Hayakawa, A. Himeno, R. Yasuhara, W. Boullart, E. Vecchio, T. Vandeweyer, T. Witters, D. Crotti, M. Jurczak, S. Fujii, S. Ito, Y. Kawashima, Y. Ikeda, A. Kawahara, K. Kawai, Z. Wei, S. Muraoka, K. Shimakawa, T. Mikawa, and S. Yoneda, *Tech. Dig. Symp. VLSI Technol.*, T14–T15 (2015).
4. K. Szot, W. Speier, G. Bihlmayer, and R. Waser, *Nat. Mater.*, **5**, 312–320 (2006).
5. K. Szot, R. Dittmann, W. Speier, and R. Waser, *Phys. Stat. Sol. (RRL)*, **1**, R86–R88 (2007).
6. J. Y. Son and Y.-H. Shin, *Appl. Phys. Lett.*, **92**, 222106 (2008).
7. C. Yoshida, K. Kinoshita, T. Yamasaki, and Y. Sugiyama, *Appl. Phys. Lett.*, **93**, 042106 (2008).
8. D.-S. Shang, L. Shi, J.-R. Sun, and B.-G. Shen, *Nanotechnology*, **22**, 254008 (2011).
9. M. Lanza, K. Zhang, M. Porti, M. Nafría, Z. Y. Shen, L. F. Liu, J. F. Kang, D. Gilmer, and G. Bersuker, *Appl. Phys. Lett.*, **100**, 123508 (2012).
10. Y. Yang, X. Zhang, L. Qin, Q. Zeng, X. Qiu, and R. Huang, *Nat. Commun.*, **8**, 15173 (2017).
11. A. K. Singh, S. Blonkowski, and M. Kogelschatz, *J. Appl. Phys.*, **124**, 014501 (2018).
12. Q. Chen, G. Liu, W. Xue, J. Shang, S. Gao, X. Yi, Y. Lu, X. Chen, M. Tang, X. Zheng, and R.-W. Li, *ACS Appl. Electron. Mater.*, **1**, 789–798 (2019).
13. G. Di Martino, A. Demetriadou, W. Li, D. Kos, B. Zhu, X. Wang, B. de Nijs, H. Wang, J. MacManus-Driscoll, and J. J. Baumberg, *Nat. Electron.*, **3**, 687–693 (2020).
14. A. Tao, T. Yao, Y. Jiang, L. Yang, X. Yan, H. Ohta, Y. Ikuhara, C. Chen, H. Ye, and X. Ma, *Nano Lett.*, **21**, 5586–5592 (2021).
15. A. Wedig, M. Luebben, D.-Y. Cho, M. Moors, K. Skaja, V. Rana, T. Hasegawa, K. K. Adepalli, B. Yildiz, R. Waser, and I. Valov, *Nat. Nanotech.*, **11**, 67–74 (2016).
16. M. Moors, K. K. Adepalli, Q. Lu, A. Wedig, C. Bäumer, K. Skaja, B. Arndt, H. L. Tuller, R. Dittmann, R. Waser, B. Yildiz, and I. Valov, *ACS Nano*, **10**, 1481–1492 (2016).
17. Q. Xu, Y. Ma, and M. Skowronski, *J. Appl. Phys.*, **127**, 055107 (2020).
18. B. A. Movchan and A. V. Demchishin, *Fiz. Met. Metalloved.*, **28**, 653–660 (1969).
19. B. Shin and M. J. Aziz, *Phys. Rev. B*, **76**, 085431 (2007).
20. A. Tsurumaki-Fukuchi, Y. Tsuta, M. Arita, and Y. Takahashi, *Phys. Stat. Sol. (RRL)*, **13**, 1900136 (2019).
21. M. A. Alam, B. E. Weir, and P. J. Silverman, *IEEE. Trans. Electron Devices*, **49**, 232–238 (2002).

22. S. Yu, Y. Wu, R. Jeyasingh, D. Kuzum, and H.-S. P. Wong, *IEEE. Trans. Electron Devices*, **58**, 2729–2737 (2011).
23. Y. Ma, D. Li, A. A. Herzing, D. A. Cullen, B. T. Sneed, K. L. More, N. T. Nuhfer, J. A. Bain, and M. Skowronski, *ACS Appl. Mater. Interfaces*, **10**, 23187–23197 (2018).
24. R. Nakamura, T. Toda, S. Tsukui, M. Tane, M. Ishimaru, T. Suzuki, and H. Nakajima, *J. Appl. Phys.*, **116**, 033504 (2014).
25. R. D. Shannon, *Acta Crystallogr.*, A**32**, 751–767 (1976).
26. D. B. Strukov, F. Alibart, and R. S. Williams, *Appl. Phys. A*, **107**, 509–518 (2012).

Chapter 5

Novel Materials and Characterization 1

Impact of Boron Doping and H_2 Annealing on Light Emission from Ge/Si Core-Shell Quantum Dots

Seiichi Miyazaki, and Katsunori Makihara

Graduate School of Engineering, Nagoya University
Furo-cho, Chikusa-ku, Nagoya 464-8603, Japan
Phone: +81-52-789-3588
E-mail: miyazaki@nuee.nagoya-u.ac.jp

We have fabricated Si-QDs with boron-doped Ge core with an areal density as high as $\sim 10^{11}$ cm^{-2} on ultrathin SiO$_2$ and characterized their room temperature photoluminescence (PL) properties. With B doping to the Ge core, stable PL observed in the energy range of 0.62 - 0.85 eV was increased by a factor of ~ 1.5. Note that a new relatively-narrow component peaked at ~ 0.64 eV emerges in addition to four components derived from the spectral deconvolution of the PL spectrum of undoped QDs and is attributable to radiative recombination between the first quantized state in the conduction band of the Si clad and the B-acceptor level in the Ge core. With H_2 post-anneal at 100°C, the integrated PL intensity of the B-doped QDs was reduced by $\sim 35\%$ and the 0.64 eV component disappeared. With increasing H_2 post-anneal temperature up to 350°C, the PL intensity was increased to ~ 1.4 times the initial intensity before post-annealing, and the 0.64 eV component was recovered. The observed PL changes with post-anneal are attributable to hydrogen-induced passivation of B acceptors at 100°C and thermal dissociation of B-H complex at higher temperatures up to 350°C accompanied with hydrogen passivation of residual dangling bond defects as seen in the undoped case.

Introduction

Light emission from Si/Ge based nanostructures has attracted much attention in the field of Si-based photonics because of its potential to combine photonic processing with electronic processing on a single chip [1-9]. In previous works, we have reported high-density formation of Ge-core/Si-clad quantum dots (QDs) by controlling thermal decomposition of GeH$_4$ and SiH$_4$ alternately on thermally-grown SiO$_2$, and demonstrated that stable room-temperature photoluminescence (PL) in the energy region from 0.66 to 0.88 eV, which was obtained with 976 nm photoexcitation of a single layer of highly-dense QDs consisting of ~ 6.0 nm Ge-core and ~ 3.0 nm-thick Si-clad in average size, can be deconvoluted into four Gaussian components originating from radiative recombination through quantized states in the QDs as verified from dot size and temperature dependences of PL properties [10-16]. A possible way to enhance the radiative recombination rate in photoexcited QDs is considered to increase electronic states assisting radiative transition with impurity doping into the QDs and reduce residual non-radiative centers if any. In this work, we have studied room temperature PL properties of Si-QDs with B-doped Ge core in comparison to those from undoped QDs. In addition, we also evaluated the impact of post-anneal in either He or N_2 ambience on the PL properties.

Experimental

After conventional wet-chemical cleaning steps of n-Si(100) substates, a ~2.0 nm-thick SiO_2 layer was grown by dry O_2 oxidation at 1000°C and followed by dipping shortly into a 0.1% HF solution to obtain surface termination uniformly with OH bonds acting as reactive sites for Si nucleation on SiO_2 [17]. After that, hemispherical Si-QDs were formed by low pressure chemical vapor deposition (LPCVD) using pure SiH_4 at 600°C and 70 Pa. Subsequently, highly-selective deposition of Ge to be core was made on pre-grown Si-QDs by LPCVD using 5% GeH_4 diluted with H_2 at 500°C, with keeping the gas pressure as low as ~27 Pa. During the Ge deposition, boron δ-doping was carried out by short pulse injection of 1% B_2H_6 diluted with He. And then, the Si-cap formation on pre-grown QDs was performed by selective LPCVD using 5% SiH_4 diluted with H_2 at 580°C and 90 Pa. Finally, for the surface passivation, the QDs surface was oxidized to form a ~1.0 nm-thick SiO_2 layer by exposing to a remote O_2/Ar plasma generated with a 60 MHz power generator at a substrate temperature of 500°C. We also fabricated undoped Si-QDs with Ge core as a reference with the same process steps except the B doping. Formation of the Si-QDs with Ge core was confirmed by atomic force microscopy (AFM) images. Photoluminescence measurements were carried out at room temperature with ~976 nm excitation using a semiconductor laser.

Results and Discussion

AFM topographic images taken after each deposition step and corresponding dot height distributions evaluated from the AFM images confirm the formation of Si-QDs with an areal density as high as ~2×10^{11} cm^{-2}, and the average size in height was determined to be ~3.0 nm in the pre-grown Si-QDs, ~3.3 nm in the Ge core, and ~2.5 nm in the Si-cap by fitting a log-normal function to each of measured size distributions as seen in Fig. 1. We have also confirmed no significant change in the dot size with B-doping. To evaluate boron activation into the Ge cores, surface potential changes before and after bias application were measured by means of an AFM/Kelvin probe technique using a Rh-coated AFM cantilever [18]. Without any bias applied to the sample surfaces, a uniform surface potential image was obtained irrespective of B-doping. When the B-doped QDs sample surface was scanned in contact mode with an AFM tip biased at 0 V with respect to the substrate, a decrease of the surface potential of the scanned area by —18 mV was detected in a non-contact Kelvin probe mode, as shown in Fig. 2 (a). But in contrast, in the cases of thermally-grown SiO_2 and undoped Si-QDs with Ge core, no changes in surface potential were detectable under the same tip-bias condition (Fig. 2 (b)). In assuming a simple equivalent circuit for surface potential measurements using the Kelvin probe method as described in Ref. 19, the measured decrease in the surface potential in Fig. 2 (a) was estimated to correspond to single electron charging to each of the Si-QDs with B-doped Ge core. Notice that, to obtain the surface potential change of ~-18 mV on the undoped Si-QDs, the tip bias for scanning in a tapping mode should be set up to −1.2V. The result of Fig. 2 (a) can be interpreted in terms of electron injection into the valence band of the B-doped Ge core, namely a negatively-charged QDs with an ionized B acceptor, as seen in the inset of the energy band diagram. In fact, with increasing negative tip-bias up to −1.2 V for B-doped QDs, negative charging is further promoted due to electron injection to the conduction band of QDs because the Fermi level of Rh lies energetically close to the conduction band edge of the Si clad with such a negative tip-bias application.

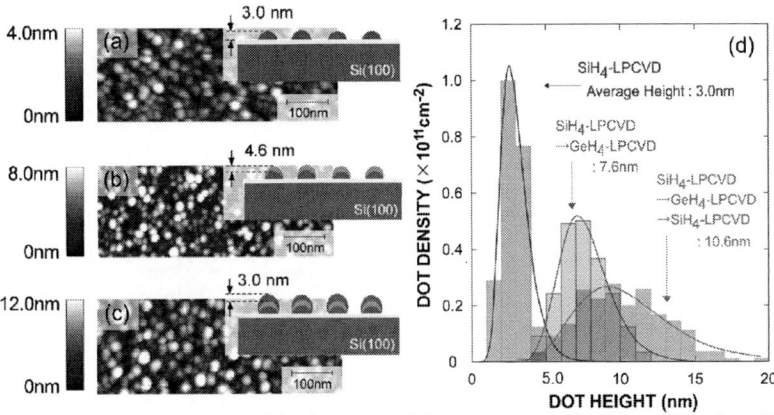

Fig. 1   Typical AFM topographic images of (a) pre-grown Si-QDs, (b) after Ge deposition, and (c) after Si-cap formation, and (d) dot height distributions evaluated from the AFM images.

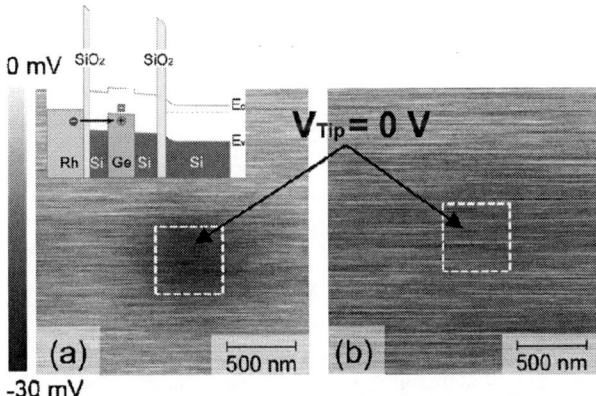

Fig. 2   Surface potential images of Si-QDs with B-doped (a) and undoped Ge core (b) measured by a AFM/Kelvin probe mode after tapping a Rh-coated AFM tip at a zero bias with respect to the substrate.   In each image, the tapping area is surrounded by a dashed line.   Energy band diagram of the sample contacted with the Rh tip at 0 V is also shown in the inset.

When the Si-QDs with undoped and B-doped Ge core were excited by a 976 nm light at a power density of 0.33 W/cm^2, stable PL signals were detected at room temperature in the energy region from 0.62 to 0.86 eV as represented in Fig. 3. As reported in Ref. 13, the observed PL spectrum of the undoped Si-QDs with Ge core can be deconvoluted into mainly four components originating from the radiative recombination of photogenerated electron-hole pairs through quantized states of the Si-QDs with Ge core. With B doping to the Ge cores, the PL intensity was increased by a factor of 1.5 compared with that of the undoped QDs, which implies the contribution of thermally-generated holes in B-doped Ge cores to the radiative recombination. It should be noted that, for the PL from the Si-QDs with B-doped Ge core (Fig. 3 (b)), another relatively-narrow component (Comp. 5) peaked at ~0.64 eV appears in addition to the four components derived from the spectral deconvolution of the PL from undoped QDs. Since the Comp 1. has been assigned to the radiative transition by coupling between electron quantized state in Si clad and hole quantized state in Ge core [13], the Comp. 5 is attributable to radiative transition from the first quantized state for electrons in the Si clad to the B-acceptor level in the Ge core.

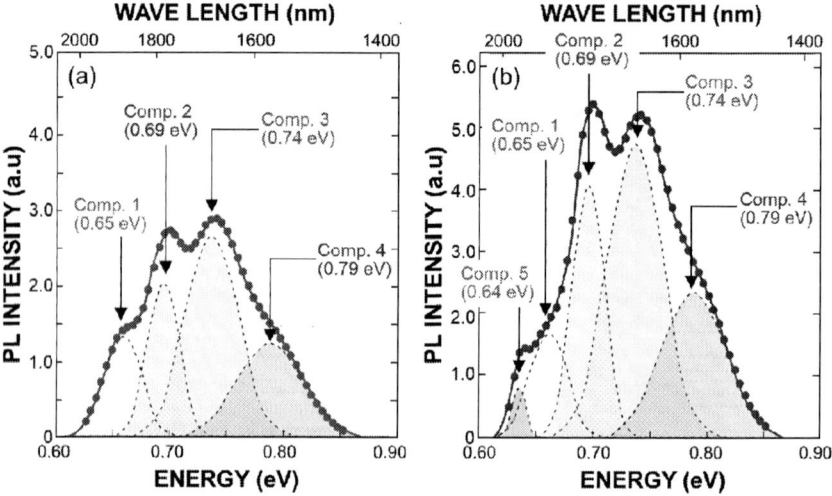

Fig. 3 Room temperature PL spectra of Si-QDs with (a) undoped and (b) B-doped Ge core, which can be deconvoluted into 4 and 5 Gaussian components, respectively, as shown in dashed lines.

We also evaluated the effect of post-annealing on light emission properties, where the Si-QDs with undoped and B-doped Ge core were annealed in either pure H$_2$ or N$_2$ in a temperature range from 100 to 600°C, as shown in the Figs. 4 and 5. In the H$_2$ post-anneal of Si-QDs with undoped Ge, with an increase in the anneal temperature up to 350°C, the integrated PL intensity was increased by a factor of 1.3 with almost no significant change in the spectral shape, namely, with keeping both the peak position and the spectral width in each deconvoluted component. In contrast, with N$_2$ post-anneal, almost no change in the PL intensity was detectable up to 350°C. Observed increase in PL intensity with the H_2 post-anneal is likely to be attributable to hydrogen passivation of residual defects in the Ge core and at the Si-clad/Ge-core interface. Further increase in the anneal temperature up to 600°C caused a significant reduction in the integrated PL intensity mainly due to loss of hydrogen passivation of dangling-bond defects at temperatures over 400°C and partly due to the defect generation in compositional mixing at Si/Ge interfaces slightly occurring at 600°C.

In post-annealing the B-doped QDs, the PL intensity was reduced by 100°C anneal regardless of H_2 or N_2 ambience, and then increased by anneal up to 350°C as seen in the undoped QDs. Figure 6 shows the integrated PL intensity of each deconvoluted component as a function of anneal temperature. In the N_2 post-anneal, no significant anneal temperature dependence was observed for the Comp. 5 concerning the B-acceptor level in the Ge cores. It is interesting to note that the Comp. 5 was hardly detectable by 100°C anneal in H_2 ambience and recovered by 350°C anneal. The result can be explained as follows. Hydrogen induced passivation of B acceptors in Ge cores during post-annealing at 100°C, that is the formation of B-H complex [20], causes weakening the PL intensity which was enhanced with B-doping in as-prepared sample. And, at temperatures of 200°C and over, thermal dissociation of the B-H complex is promoted and results in recovery of the B-doping effect on the PL intensity accompanied with dangling-bond defect passivation. The observed changes in PL intensity with N_2 anneal is presumably due to the effect of residual hydrogen atoms which incorporate interstitially in the dots in δ-doping using a B_2H_6 gas during Ge core formation. The bonding change in hydrogen to generate bangling bond defects and out diffusion of interstitial hydrogen promoting at temperatures below 200°C may be involved in the observed reduction of the PL intensity in N_2 anneal.

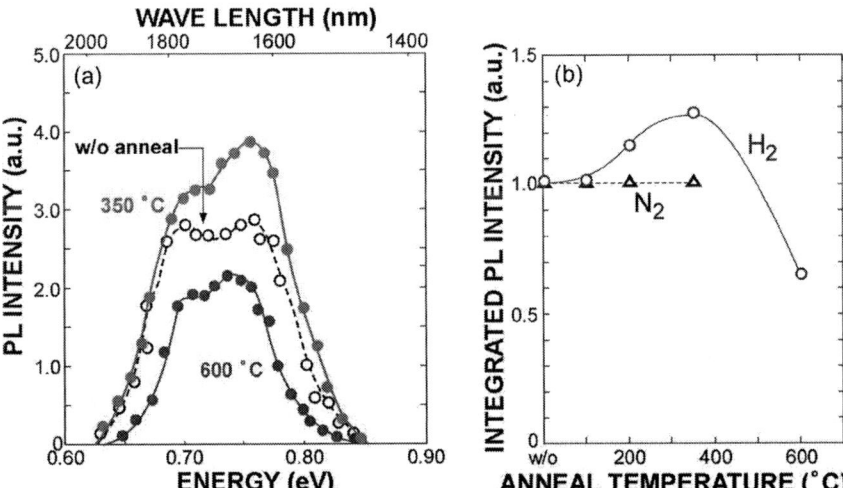

Fig. 4   Room temperature PL spectra of Si-QDs with undoped Ge core before and after H_2 post-anneal at 350 and 600°C (a) and integrated PL intensities plotted as functions of anneal temperature in H_2 and N_2 ambience (b).

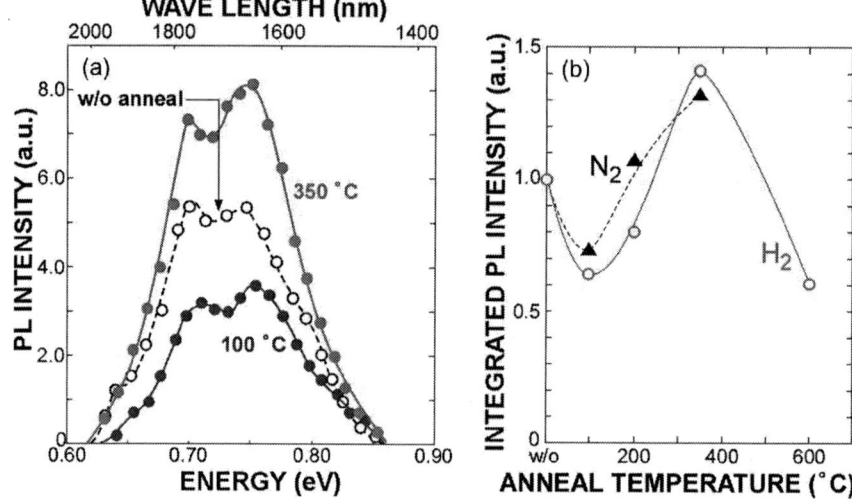

Fig. 5 Room temperature PL spectra of Si-QDs with B-doped Ge core before and after H_2 post-anneal at 100 and 350°C (a) and integrated PL intensities plotted as functions of anneal temperature in H_2 and N_2 ambience (b).

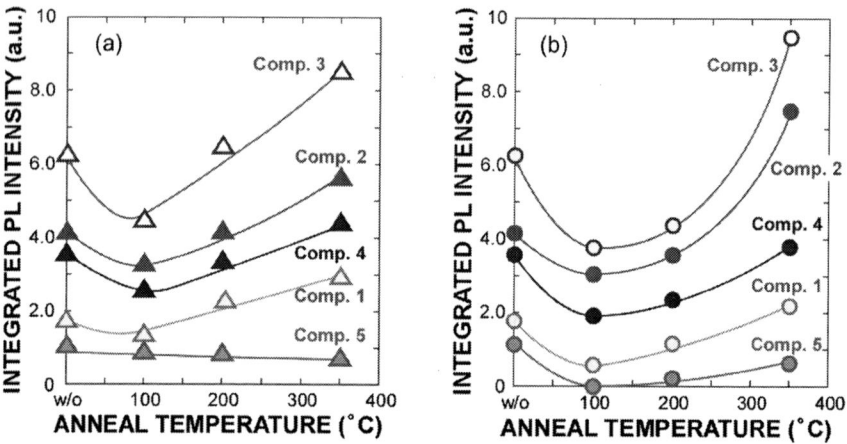

Fig. 6 Integrated intensities of each PL component for the Si-QDs with B-doped Ge core, evaluated from the spectral analysis using a Gaussian curve fitting method, as functions of anneal temperature in N_2 (a) and H_2 (b) ambience.

Conclusions

We have demonstrate that boron doping to Ge core in Si-QDs is effective to enhance near-infrared photoluminescence (PL) under ~976 nm excitation at room temperature with advent of a relatively-narrow PL component peaked at ~0.64 eV being attributable to radiative recombination concerning B acceptor levels. From a comparison of PL intensities from the QDs with and without B-doping in the Ge core, it is suggested that holes confined in the Ge core play a role in enhancing the electron-hole recombination rate. We also found that H_2 post-anneal at 350°C on the Si-QDs with B-doped Ge core is an efficient way to improve PL efficiency presumably by hydrogen passivation of residual dangling-bond defects as well as B activation due to thermal dissociation of B-H complex.

Acknowledgment

This work was supported in part by Grant-in Aids for Scientific Research (A) No. 19H00762 from the Ministry of Education, Culture, Sports, Science and Technology, Japan.

References

1. G. F. Grom, D. J. Lockwood, J. P. McCaffrey, H. J. Labbé, P. M. Fauchet, B. White, Jr, J. Diener, D. Kovalev, F. Koch, and L. Tsybeskov, *Nature* **407**, 358–361 (2000).
2. R. J. Walters, G. I. Bourianoff, and H. A. Atwater, *Nature Mat.* **4**, 143–146 (2005).
3. O. Jambois, H. Rinnert, X. Devaux, and M. Vergnat, *J. Appl. Phys.* **98**, 046105 (2005).
4. K. W. Sun, S. H. Sue, and C. W. Liu, *Physica E: Low-dimensional Systems and Nanostructures* **28**, 525 (2005).
5. A Perez del Pino, E. Gyorgy, I. C. Marcus, J. Roqueta, and M. I. Alonso, *Nanotechnology* **22**, 295304 (2011).
6. X. Xu, N. Usami, T. Maruizumi, and Y. Shiraki, *J. Crystal Growth* **378**, 636–639 (2013).
7. K. Nishida, X. Xu, K. Sawano, T. Maruizumi, and Y. Shiraki, *Thin Solid Films* **557**, 66–69 (2014).
8. B. Dutt, D. S. Sukhdeo, D. Nam, B. M. Vulovic, Z. Yuan, K. C. Saraswat, *IEEE Photonics Journal* **4**, 2002–2009 (2012)
9. S. Wirths, R. Geiger, N. von den Driesch, G. Mussler, T. Stoica, S. Mantl, Z. Ikonic, M. Luysberg, S. Chiussi, J. M. Hartmann, H. Sigg, J. Faist, D. Buca, and D. Grützmacher, *Nature Photonics* **9**, 88–92 (2015).
10. Y. Darma, R. Takaoka, H. Murakami, and S. Miyazaki, *Nanotechnology* **14**, 413–415 (2003).
11. K. Makihara, K. Kondo, M. Ikeda, A. Ohta, and S. Miyazaki, *ECS Transactions* **64**, 365–370 (2014).
12. K. Yamada, K. Kondo, K. Makihara, M. Ikeda, A. Ohta, and S. Miyazaki, *ECS Transactions* **75**, 695–700 (2016).

13. K. Kondo, K. Makihara, M. Ikeda, and S. Miyazaki, *J. Applied Physics* **119**, 033103 (2016).
14. S. Miyazaki, K. Makihara, A. Ohta, and M. Ikeda, *Technical Digest of Int. Electron Devices Meeting 2016*, 826-830 (2016).
15. S. Miyazaki, K. Yamada, K. Makihara, and M. Ikeda, *ECS Transactions* **80**, 167-172 (2017).
16. K. Makihara, M. Ikeda, N. Fujimura, K. Yamada, A. Ohta, and S. Miyazaki, *Applied Physics Express* **11**, 011305 (2018).
17. S. Miyazaki, Y. Hamamoto, E. Yoshida, M. Ikeda, and M. Hirose, *Thin Solid Films* **369**, 55–59 (2000).
18. N. Shimizu, M. Ikeda, E. Yoshida, H. Murakami, S. Miyazaki, and M. Hirose, *Jpn. J. Appl. Phys.* **39**, 2318–2320 (2000).
19. K. Makihara, and S. Miyazaki, *Jpn. J. Appl. Phys.* **49**, 065002 (2010).
20. M. Stavola, *Acta Physica Polanica A* **82**, 585-598 (1992).

Study of HfO$_2$-based High-k gate Insulators for GaN Power Device

T. Nabatame[a], E. Maeda[a], M. Inoue[a], M. Hirose[a], R. Ochi[b], T. Sawada[a], Y. Irokawa[a], T. Hashizume[b], K. Shiozaki[c], T Onaya[d,e], K. Tsukagoshi[a], and Y. Koide[a]

[a] National Institute for Materials Science, Tsukuba, 305-0044, Japan
[b] Research Center for Integrated Quantum Electronics (RCIQE), Hokkaido University, Sapporo 060-0813, Japan
[c] Institute of Materials and Systems for Sustainability, Nagoya University, Nagoya, 464-8601, Japan
[d] National Institute of Advanced Industrial Science and Technology, Tsukuba, 305-8568, Japan
[e] JSPS Research Fellow PD Tokyo 102-0083, Japan

We discussed about usefulness of HfO$_2$-based high dielectric constant (High-k) materials such as HfO$_2$, HfSiO$_x$ and HfAlO$_x$ as gate insulator for GaN power device. Here, we systematically studied characteristics of Hf$_{0.55}$Al$_{0.45}$O$_x$ gate insulator which fabricated by plasma-enhanced atomic layer deposition and post-deposition annealing (PDA) at 800°C in N$_2$ ambient. The Hf$_{0.55}$Al$_{0.45}$O$_x$ film had an amorphous structure but Ga diffusion into Hf$_{0.55}$Al$_{0.45}$O$_x$ film was observed after PDA. The k-value of the Hf$_{0.55}$Al$_{0.45}$O$_x$ film was 17.2, which was larger than that of HfSiO$_x$ (13.5). The Hf$_{0.55}$Al$_{0.45}$O$_x$ also showed superior electrical properties such as a minimal flatband voltage (V_{fb}) hysteresis (+10 mV) and a relatively small V_{fb} shift (-0.95 V), as well as a low interface state density (~ 1×10^{11} cm^{-2}eV^{-1} at E_c-E=0.25 eV), and a high breakdown electric field (8.6 MVcm^{-1}). Based on these experimental data and previous our research of HfO$_2$ and HfSiO$_x$ films, we concluded the HfSiO$_x$ and HfAlO$_x$ had candidate materials as gate insulator for GaN power device.

Introduction

Metal oxide materials with high dielectric constant (High-k) have widely been investigated as gate insulator for vertical GaN metal-oxide-semiconductor (MOS) device and high-electron mobility transistor (HEMT) with high-power and high-frequency performance. High-k gate insulator was expected to improve characteristics such as reduction of a leakage current density (J) at high electric field (E) and a high breakdown electric field (E_{BD}) according to the increase of physical thickness due to a high k-value. Numerous High-k gate insulators, including Al$_2$O$_3$ (1-11), AlON (12, 13), ZrO$_2$ (14, 15), HfO$_2$ (14, 16-18), SiO$_2$-HfO$_2$ (19), Al$_2$O$_3$-HfO$_2$ (20), and Al$_2$O$_3$-TiO$_2$ bilayers (21), (Al,Si)O (22) AlSiO$_x$ (23-25) and HfSiO$_x$ (16, 18, 26, 27) have been studied for use in MOS structures, and films of these materials are typically deposited via chemical vapor deposition or atomic layer deposition (ALD). ALD is the optimal approach to fabricate conformal films on three-dimensional structures, which is required when producing vertical GaN devices. In addition, to improve characteristics of High-k gate insulators, post-deposition annealing (PDA) process was generally carried out after High-k deposition. We have also been studied High-k gate insulators such as HfO_2 and $HfSiO_x$ using n-GaN/High-k/Pt MOS capacitors (16, 18, 26).

The HfO_2 film began to crystallize in the low temperature region of 400°C and had a monoclinic phase (28). Although the HfO_2 film had higher k-value of 17.6, a small E_{BD} of 3.0 MVcm^{-1}, a high flatband voltage (V_{fb}) hysteresis of +600 mV, a high V_{fb} shift (+0.97 V) and a high interface density (D_{it}) of ~ 3×10^{12}cm^{-2}eV^{-1} at E_c-E=0.25eV (16, 18).

The $HfSiO_x$ film with Hf-rich composition had an amorphous structure even after PDA at 800°C in N_2. The $HfSiO_x$ film exhibited a high k-value of 13.5, a high E_{BD} of 8.6 MVcm^{-1}, no frequency dispersion, a small V_{fb} hysteresis of $\leq$+70 mV, a high V_{fb} shift (-0.45V) and a small D_{it} of ~ 6×10^{11} cm^{-2}eV^{-1} at E_c-E=0.25eV (16, 18, 26).

Here, we pay attention to the $HfAlO_x$ film as gate insulator of GaN power device because of a high potential material including a high k-value, a relatively high bandgap, and stable amorphous structure in the high temperature region (28, 29).

In this study, we systematically investigated characteristics of Pt-gated metal-oxide-semiconductor (MOS) capacitors with Hf-rich $Hf_{0.55}Al_{0.45}O_x$ gate insulators. We also discuss usefulness of HfO_2, $HfSiO_x$ and $HfAlO_x$ as gate insulator for GaN power device according to our previous research of the HfO_2, and $HfSiO_x$ films.

Experimental

The fabrication flow of Pt-gated n-GaN MOS capacitors with $Hf_{0.55}Al_{0.45}O_x$ gate insulator is shown in **figure 1**. The n$^+$-GaN (0001) substrates (N_d: 1.3 $\times 10^{18}$ cm^{-3}) with 5-μm-thick Si-doped GaN epilayers (N_d: 2 $\times 10^{16}$ cm^{-3}) were firstly cleaned with a sulfuric acid/peroxide mixture to remove organic residues and subsequently cleaned with buffered hydrofluoric acid. Next, $(HfO_2)_m(Al_2O_3)_n$ laminate was deposited on n-GaN epilayer by plasma-enhanced atomic layer deposition (PE-ALD) at 300°C using trimethylaluminum

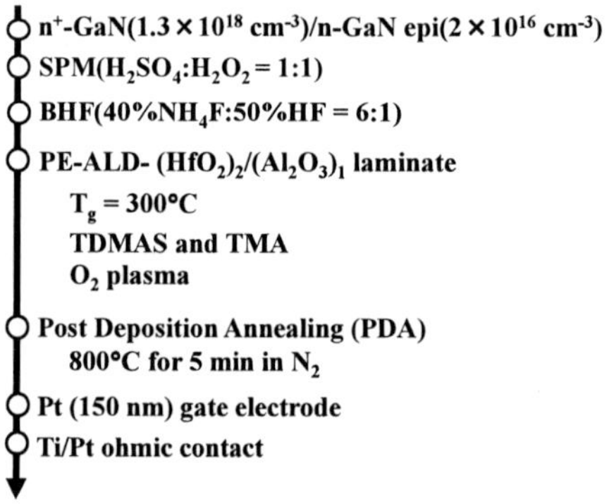

n$^+$-GaN(1.3 × 10^{18} cm^{-3})/n-GaN epi(2 × 10^{16} cm^{-3})

SPM(H$_2$SO$_4$:H$_2$O$_2$ = 1:1)

BHF(40%NH$_4$F:50%HF = 6:1)

PE-ALD- (HfO$_2$)$_2$/(Al$_2$O$_3$)$_1$ laminate
  T$_g$ = 300°C
  TDMAS and TMA
  O$_2$ plasma

Post Deposition Annealing (PDA)
  800°C for 5 min in N$_2$

Pt (150 nm) gate electrode

Ti/Pt ohmic contact

Figure 1. Fabrication process flow of n-GaN/Hf$_{0.55}$Al$_{0.45}$O$_x$/Pt MOS capacitors.

and tetrakis(dimethylamino)hafnium precursors and plasma oxygen gases. The Hf proportion in the $(HfO_2)_m(Al_2O_3)_n$ laminate was varied by changing the numbers of each ALD cycle of the HfO_2 and Al_2O_3 (m/n=2/1) used to deposit the HfO_2 and Al_2O_3 layers as shown in **figure 2**. The $Hf_{0.55}Al_{0.45}O_x$ film was formed from $(HfO_2)_2(Al_2O_3)_1$ laminate after PDA at 800°C in N_2. Finally, 100-nm-thick Pt gate electrodes were deposited on the $HfAlO_x$ gate insulators through a shadow mask and Pt(100 nm)/Ti(20 nm) ohmic contacts were deposited on the n^+-GaN substrate.

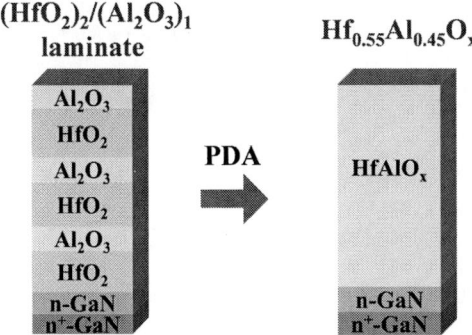

Figure 2. Schematic illustrations of $Hf_{0.55}Al_{0.45}O_x$ formation from $(HfO_2)_2(Al_2O_3)_1$ laminate by PDA at 800°C in N_2.

Current-voltage measurements were performed at room temperature under dark condition using a Keithley 4200-SCS semiconductor characterization system. Capacitance-voltage (C-V) measurements were performed at room temperature under dark condition, using an Agilent B1500A semiconductor device parameter analyzer. During these measurements, the gate bias was swept from the depletion region to the accumulation region and back to the depletion at a frequency ranging from 1 MHz to 1 kHz. The V_{fb} values for each specimen were estimated from the C–V data using a MIRAI-ACCEPT software (30, 31).

The extent of crystallization and the morphology of the $HfAlO_x$ films were examined using transmission electron microscopy (TEM) and energy dispersive X-ray spectroscopy (EDS). The Ga depth profiles in the $HfAlO_x$ film before (as-grown) and after PDA were evaluated by secondary ion mass spectrometry (SIMS).

Results and discussion

The Hf:Al ratios in the $(HfO_2)_2(Al_2O_3)_1$ laminate made with ALD cycle ratios of m/n=2/1 was estimated to be 0.55:0.45 ($Hf_{0.55}Al_{0.45}O_x$) based on EDS data. A cross-sectional TEM image of n-GaN/$Hf_{0.55}Ali_{0.45}O_x$ film after PDA at 800°C is shown in **figure 3**. This $Hf_{0.55}Ali_{0.45}O_x$ film had an amorphous structure because the electron diffraction pattern exhibited no ring or spot-like features. The thickness was about 25 nm. In previous our research, we found an epitaxial crystalline Ga_2O_3 with a thickness of ~1.5 nm was grown between SiO_2 gate insulator and GaN (32). However, it is very difficult to observe the existence of Ga_2O_3 interfacial layer at n–GaN/$Hf_{0.55}Al_{0.45}O_x$ interface because Ga_2O_3 is composed of lattice-matched hexagonal crystal on the surface of n-GaN with the same structure.

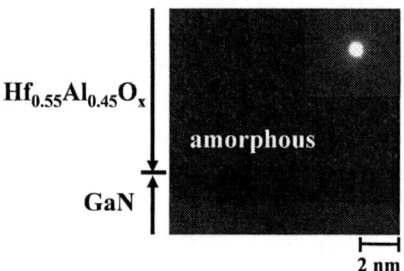

Figure 3. A cross-sectional TEM image and the electron diffraction pattern of GaN/$Hf_{0.55}Al_{0.45}O_x$ film after PDA at 800°C in N_2.

Figure 4 shows the Ga depth profile of the n-GaN/$Hf_{0.55}Al_{0.45}O_x$ film before (as-grown) and after PDA at 800°C. The Ga concentration of the as-grown $Hf_{0.55}Al_{0.45}O_x$ film was close to the detection limit in the depth region between 5-20 nm. The Ga diffusion of the PDA $Hf_{0.55}Al_{0.45}O_x$ film was observed and the Ga concentration was at least an order of magnitude larger than the as-grown $Hf_{0.55}Al_{0.45}O_x$ film. Furthermore, Ga diffusion profile near interface between n-GaN and $Hf_{0.55}Al_{0.45}O_x$ film was found to satisfy Fick's law. The Ga diffusion of the $Hf_{0.55}Al_{0.45}O_x$ film was faster than that of the $HfSiO_x$ in the same PDA temperature of 800°C (not shown).

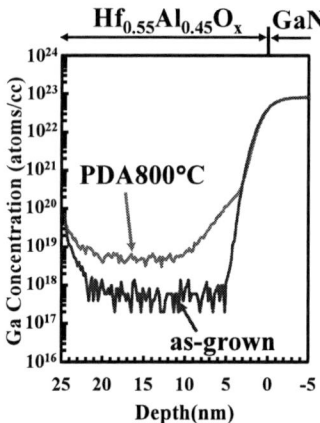

Figure 4. The SIMS for Ga depth profile of the n-GaN/$Hf_{0.55}Al_{0.45}O_x$ film before (as-grown) and after PDA at 800°C.

Figure 5 presents the C-V characteristics of the $Hf_{0.55}Al_{0.45}O_x$ capacitor at measurement frequencies ranging from 1 MHz to 1 kHz. Note that the $Hf_{0.55}Al_{0.45}O_x$ capacitor exhibited a small degree of V_{fb} hysteresis of +10 mV and no frequency dispersion under the depletion bias conditions. In previous our research, the HfSiO$_x$ capacitor exhibited a similar V$_{fb}$ hysteresis and no frequency dispersion, but the HfO$_2$ capacitor showed a significant V$_{fb}$ hysteresis (around +600 mV) at all frequencies (16, 18). The V$_{fb}$ shift from an ideal V$_{fb}$ was -0.95 V in the Hf$_{0.55}$Al$_{0.45}$O$_x$ capacitor. The k-value which estimated from saturated capacitance, of the Hf$_{0.55}$Al$_{0.45}$O$_x$ capacitor was 17.2, which was higher than the HfSiO$_x$ capacitor (13.5).

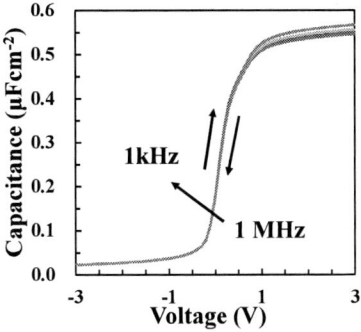

Figure 5. the C-V characteristics of the Hf$_{0.55}$Al$_{0.45}$O$_x$ capacitor

To better understand the characteristics of the n-GaN/Hf$_{0.55}$Al$_{0.45}$O$_x$ interfaces, the D$_{it}$ was estimated using a conductance method previously developed by Nicollian and Goetzberger (33). An equivalent parallel conductance (G$_p$/ω) in depletion bias condition is calculated as the following Eq. (1):

$$\frac{G_p}{\omega} = \frac{\omega C_{ox}^2 G_m}{G_m^2 + \omega^2 (C_{ox} - C_m)^2} \quad \text{Eq. (1)}$$

Here, ω, C$_{ox}$, G$_m$ and C$_m$ are frequency, oxide capacitance, measured conductance and measured capacitance, respectively. The energy distributions of the estimated D$_{it}$ values for the Hf$_{0.55}$Al$_{0.45}$O$_x$ capacitor is shown in **figure 6**. The Hf$_{0.55}$Al$_{0.45}$O$_x$ capacitor exhibited

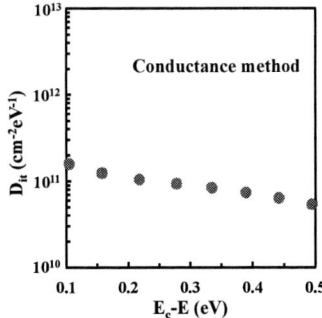

Figure 6. The D$_{it}$ energy distributions for the Hf$_{0.55}$Al$_{0.45}$O$_x$ capacitor

significantly reduced D_{it} values, as low as $\sim 1 \times 10^{11}$ cm^{-2}eV^{-1} at an energy level of E_c-E=0.25 eV compared to the HfO$_2$ capacitor of $\sim 3 \times 10^{12}$cm^{-2}eV^{-1} and the HfSiO$_x$ capacitor of $\sim 6 \times 10^{11}$ cm^{-2}eV^{-1}. The energy level of E_c - E was selected to be 0.25 eV corresponding to E1 electron traps, respectively, where E_c is the energy level of the conduction band minimum (34). Therefore, it is thought that the interface property between GaN and Hf$_{0.55}$Al$_{0.45}$O$_x$ film has less electrical defects than the HfO$_2$ and HfSiO$_x$ films.

Figure 7 shows the leakage current density (J) - electric field (E) characteristics of the Hf$_{0.55}$Al$_{0.45}$O$_x$ capacitor. The E was calculated from the applied voltage divided by the physical thickness of the Hf$_{0.55}$Al$_{0.45}$O$_x$ film. The J behavior of the Hf$_{0.55}$Al$_{0.45}$O$_x$ capacitor was very similar to the HfSiO$_x$ capacitor. On the other hand, the HfO$_2$ capacitor exhibited a spike-loke leakage current because of polycrystalline structure (16, 18). The breakdown electric field (E_{bd}) values, defined at a current density of 1×10^1 A/cm^2, for the Hf$_{0.55}$Al$_{0.45}$O$_x$ capacitor was 8.6 MV/cm, which was the same value to the HfSiO$_x$ capacitor. In contrast, the HfO$_2$ capacitor had a much smaller E_{bd} value of 3.0 MV/cm. It is thought that the difference in the J properties between the Hf$_{0.55}$Al$_{0.45}$O$_x$ and HfO$_2$ films can be primarily attributed to the presence of amorphous or polycrystalline structures, because grain boundaries in polycrystalline structures can act current leakage paths. In fact, a polycrystalline Al$_2$O$_3$ film annealed at 900°C also exhibited a small E_{bd} value of about 5 - 7 MV/cm (8, 22).

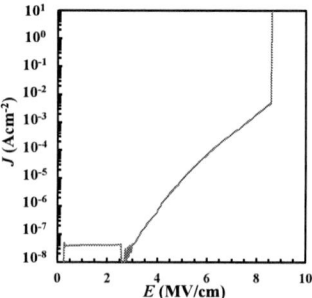

Figure 7. J - E characteristics of the Hf$_{0.55}$Al$_{0.45}$O$_x$ capacitor.

Conclusions

We systematically investigated characteristics of the Hf$_{0.55}$Al$_{0.45}$O$_x$ gate insulators, which was fabricated by PDA at 800°C in N$_2$ from the (HfO$_2$)$_2$(Al$_2$O$_3$)$_1$ laminate with PE-ALD process. The Hf$_{0.55}$Al$_{0.45}$O$_x$ had an amorphous structure. Although the Ga diffusion into the Hf$_{0.55}$Al$_{0.45}$O$_x$ film was observed, the Hf$_{0.55}$Al$_{0.45}$O$_x$ capacitors can be obtained superior electrical properties such as a high k-value of 17.2, no frequency dispersion, a small V_{fb} hysteresis of +10 mV, a small D_{it} of $\sim 1 \times 10^{11}$ cm^{-2}eV^{-1} at E_c-E=0.25 eV, a high E_{bd} value of 8.6 MV/cm. Here we compare characteristics of HfO$_2$, HfSiO$_x$ and HfAlO$_x$ as candidate material of gate insulator for GaN power device. The HfO$_2$ is firstly removed from candidate material because of poor electrical properties such as a high V_{fb} hysteresis, a high D_{it}, a small E_{bd} due to a polycrystalline structure in the same PDA temperature of 00°C. On the other hand, the $HfSiO_x$ showed superior electrical properties as well as the $HfAlO_x$. The $HfAlO_x$ had an advantage of high k (17.2) compared to the $HfSiO_x$, and it is expected to obtain high g_m in AlGaN/GaN HEMT. For $HfAlO_x$, the Ga diffusion is one concern, and it is necessary to study influence of Ga diffusion on characteristics including reliability in the future.

Acknowledgement

This work was supported by JSPS KAKENHI (Grant Nos. JP21J01667 and JP20H02189). A part of this work was also supported by the MEXT "Program for research and development of next-generation semiconductor to realize energy-saving society." Program Grant Number JPJ005357. and MEXT-Program for Creation of Innovative Core Technology for Power Electronics Grant Number JPJ009777.The authors wish to thank Dr. A. Ohi, Dr. N. Ikeda, and the members of the nanofabrication group of the National Institute for Materials Science for their support during this study.

References

1. C. Gupta, S. H. Chan, C. Lund, A. Agarwal, O. S. Koksaldi, J. Liu, Y. Enatsu, S. Keller, and U. K. Mishra, Appl. Phys. Express **9**, 121001 (2016).
2. S. Kaneki, J. Ohira, S. Toiya, Z. Yatabe, J. T. Asubar, and T. Hashizume, Appl. Phys. Lett. **109**, 162104 (2016).
3. J.-T. Asubar, Z. Yatabe, D. Cregusova, and T. Hashizume, J. Appl. Phys. **129**, 21102 (2021).
4. A. Uedono, T. Nabatame, W. Egger, T. Koschine, C. Hugenschmidt, M. Dickmann, M. Sumiya, and S. Ishibashi, J. Appl. Phys. **123**, 155302 (2018).
5. N. Taoka, T. Kubo, T. Yamada, T. Egawa, and M. Shimizu, Jpn. J. Appl. Phys. **57**, 01AD04 (2018).
6. R. D. Long, C. M. Jackson, J. Yang, A. Hazeghi, C. Hitzman, S. Majety, A. R. Arehart, Y. Nishi, T. P. Ma, S. A. Ringel, and P. C. McIntyre, Appl. Phys. Lett. **103**, 201607 (2013).
7. T. Marron, S. Takashima, Z. Li, and T. P. Chow, Phys. Status Solidi C **9**, 907 (2012).
8. K. Yoshitsugu, M. Horita, Y. Ishikawa, and Y. Uraoka, Phys. Status Solidi C **10**, 1426 (2013).
9. M. Esposto, S. Krishnamoorthy, D. N. Nath, S. Bajaj, T.-H. Hung, and S. Rajan, Appl. Phys. Lett. **99**, 133503 (2011).
10. A. J. Kerr, E. Chagarov, S. Gu, T. Kaufman-Osborn, S. Madisetti, J. Wu, P. M. Asbeck, S. Oktyabrsky, and A. C. Kummel, J. Chem. Phys. **141**, 104702 (2014).
11. K. Yuge, T. Nabatame, Y. Irokawa, A. Ohi, N. Ikeda, L. Sang, Y. Koide, and T. Ohishi, Semicond. Sci. Technol. **34**, 034001 (2019).
12. A. Uedono, T. Yamada, T. Hosoi, W. Egger, T. Koschine, C. Hugenschmidt, M. Dickmann, and H. Watanabe, Appl. Phys. Lett. **112**, 182103 (2018).
13. R. Asahara, M. Nozaki, T. Yamada, J. Ito, S. Nakazawa, M. Ishida, T. Ueda, A. Yoshigoe, T. Hosoi, T. Shimura, H. Watanabe, Appl. Phys. Express **9**, 101002 (2016).
14. K. M. Bothe, P. A. von Hauff, A. Afshar, A. Foroughi-Abari, K. C. Cadien, and D. W. Barlage, IEEE Trans. Electron Devices **60**, 4119 (2013).
15. M. Zheng, G. Zhang, X. Wang, J. Wan, H. Wu, and C. Liu, Nanoscale Research Lett. **12**, 267 (2017).
16. T. Nabatame, E. Maeda, M. Inoue, K. Yuge, M. Hirose, K. Shiozaki, M. Ikeda, T. Ohishi, and A. Ohi, Appl. Phys. Express **12**, 011009 (2019).

17. Y. C. Chang, H. C. Chiu, Y. J. Lee, M. L. Huang, K. Y. Lee, M. Hong, Y. N. Chiu, J. Kwo, and Y. H. Wang, Appl. Phys. Lett. **90**, 232904 (2007).
18. T. Nabatame, E. Maeda, M. Inoue, M. Hirose, H. Kiyono, Y. Irokawa, K. Shiozaki, and Y. Koide, ECS Trans. **92**, 109 (2019).
19. A. Taube, R. Kruszka, M. Borysiewicz, S. Gieraltowska, E. Kamińska, and A. Piotrowska, ACTA Physica Polonica A **120**, A-22 (2011).
20. J. Son, V. Chobpattana, B. M. McSkimming, and S. Stemmer, Appl. Phys. Lett. **101**, 102905 (2012).
21. D. Wei, J. H. Edgar, D. P. Briggs, S. T. Retterer, B. Srijanto, D. K. Hensley, and H. M. Meyer, J. Vac. Sci. Technol. B **32**, 060602 (2014).
22. S. H. Chan, M. Tahhan, X. Liu, D. Bisi, C. Gupta, O. Koksaldi, H. Li, T. Mates, S. P. DenBaars, S. Keller, and U. K. Mishra, Jpn. J. Appl. Phys. **55**, 021501 (2016).
23. D. Kikuta, K. Ito, T. Narita, and T. Mori, J. Vac. Sci. Technol. A **35**, 01B122 (2017).
24. D. Kikuta, K. Itoh, T. Narita, T. Kachi, Appl. Phys. Express **13**, 026504 (2020).
25. K. Ito, K. Tomita, D. Kikuta, M. Horita, T. Narita, J. Appl. Phys. **129**, 084502 (2021).
26. E. Maeda, T. Nabatame, K. Yuge, M. Hirose, M. Inoue, A. Ohi, N. Ikeda, K. Shiozaki, and H. Kiyono, Microelectron. Eng. **216**, 111036 (2019).
27. R. Ochi, E. Maeda, T. Nabatame, K. Shiozaki, T. Sato, and T. Hashizume, AIP Adv. **10**, 065215 (2020).
28. A. Toriumi, K. Iwamoto, H. Ota, M. Kadoshima, W. Mizubayashi, T. Nabatame, A. Ogawa, K. Tominaga, T. Horikawa and H. Satake, Microelectron. Eng. **80**, 190 (2005).
29. J. Robertson, and B. Falabretti, J. Appl. Phys. **100**, 014111 (2006).
30. N. Yasuda, H. Ota, T. Horikawa, T. Nabatame, H. Satake, A. Toriumi, Y. Tamura, T. Sasaki, and F. Ootsuka, The Int. Conf. on Solid State Devices and Materials, 2005, p. 250.
31. T. Nabatame, A. Ohi, T. Chikyo, M. Kimura, H. Yamada, and T. Ohishi, J. Vac. Sci. Technol. B **32**, 03D121 (2014).
32. K. Mitsuishi, K. Kimoto, Y. Irokawa, T. Suzuki, K. Yuge, T. Nabatame, S. Takashima, K. Ueno, M. Edo, K. Nakagawa, and Y. Koide, Jpn. J. Appl. Phys. **56**, 110312 (2017).
33. E. H. Nicollian, and A. Getzberger, Appl. Phys. Lett. **10** 60 (1967).
34. Y. Tokuda, ECS trans. **75**, 39 (2016).

Importance of Annealing Step on Dielectric Constant of ZrO_2 Layer of MIM Capacitors with Al_2O_3/ZrO_2 and ZrO_2/Al_2O_3 Stack Structures

T. Sawada[a], T. Nabatame[a], T. Onaya[b,c], M. Inoue[a], A. Ohi[a], N. Ikeda[a] and K. Tsukagoshi[a]

[a] International Center for Materials Nanoarchitectonics (WPI-MANA), National Institute for Materials Science (NIMS), 1-1 Namiki, Tsukuba, Ibaraki 305-0044, Japan
[b] National Institute of Advanced Industrial Science and Technology (AIST), 1-1-1 Umezono, Tsukuba, Ibaraki 305-8568, Japan
[c] JSPS Research Fellow PD, 5-3-1 Kojimachi, Chiyoda-ku, Tokyo 102-0083, Japan

We investigated characteristics of ZrO_2/Al_2O_3 (ZA) and Al_2O_3/ZrO_2 (AZ) insulators after two annealing step such as post-deposition annealing (PDA) at 600°C before TiN top electrode (TE-TiN) deposition and post-metallization annealing (PMA) at 600°C after TE-TiN deposition. The ZrO_2 layers (2.9-5.7 nm) and Al_2O_3 layer (2.0 nm) were deposited by atomic layer deposition at 300°C. ZrO_2 layer had a polycrystalline structure while Al_2O_3 layer had an amorphous structure. The capacitance of the AZ capacitors after PDA or PMA reasonably decreased as ZrO_2 thickness increased. On the other hand, the ZA capacitors after PDA (ZA wPDA) exhibited almost similar capacitance while the ZA capacitors after PMA decreased as well as the AZ capacitor. The dielectric constant (k) of the ZA wPDA drastically increased to 44 compared to other capacitors (30-37) in the ZrO_2 thickness region of over 4.9 nm. This is because the ZrO_2 layer was annealed with an asymmetric structure sandwiched between Al_2O_3 and TiN with different coefficients of thermal expansion during PDA. Furthermore, the ZA wPDA with the ZrO_2 layers (4.9-5.7 nm) exhibited superior characteristics satisfying a high k-value and a high electric filed value at $J = 1 \times 10^{-5}$ Acm^{-2}. These indicated that PDA treatment should perform after 1st-ZrO_2/Al_2O_3 layer deposition to obtain ZrO_2 layer with high k-value.

Introduction

Several stack structures such as 1st $ZrO_2/Al_2O_3/2^{nd}$ ZrO_2 (ZAZ), 1st $TiO_2/Al_2O_3/2^{nd}$ TiO_2 (TAT), and 1st $ZrO_2/(Ta/Nb)O_x$-$Al_2O_3/2^{nd}$ ZrO_2 (ZTNAZ) as insulator of metal-insulator-metal (MIM) capacitor have been widely investigated for DRAM (1-7). ZAZ is one of the most candidate structure because ZrO_2 has a high dielectric-constant (k) value (>30) and wide band gap (~5.8 eV) (1, 2). An amorphous Al_2O_3 interlayer of the ZAZ structure acts as blocking layer to suppress the leakage current flow through the grain boundaries of ZrO_2 layer. An annealing process was carried out to improve characteristics such as leakage current density (J) and k-value of the ZAZ. Two annealing processes were approached as follows: one was post-deposition annealing (PDA) after ZAZ deposition on TiN-bottom electrode (BE-TiN) and the other was post-metallization annealing (PMA) after TiN-top electrode (TE-TiN) on BE-TiN/ZAZ. However, maximum annealing temperature was limited < 650°C from the viewpoint of manufacturing process and results in a low k-value (< 39) (2). To overcome this issue, we pay attention to the k-values of the ZrO_2 layer of ZrO_2/Al_2O_3 (ZA) and Al_2O_3/ZrO_2 (AZ) stack structure assuming the 1st ZrO_2 and 2nd ZrO_2 layers of ZAZ structure.

In this study, we investigate characteristics of BE-TiN/ZA/TE-TiN and BE-TiN/AZ/TE-TiN capacitors after PDA and PMA at 600°C in N_2 and also discuss about k-value change of ZrO_2 layer of the ZA and AZ insulators.

Experimental

The fabrication process flow and schematics of MIM capacitors are shown in **figure 1**. A native oxide SiO_2 which grown on the surface of p^+-Si substrate (< 0.01 Ω/cm) was removed using buffered hydrofluoric acid solution. A 50-nm-thick BE-TiN was deposited on the p^+-Si substrate by reactive sputtering. Next, for ZA stack structure, ZrO_2 layer was deposited on BE-TiN by ALD at 300°C using $(C_5H_5)Zr[N(CH_3)_2]_3$ precursor and water gases and subsequently a 2-nm-thick Al_2O_3 layer was deposited on ZrO_2 layer by ALD at 300°C using $Al(CH_3)_3$ and water gas. The thickness of the ZrO_2 layer was varied from 2.9 to 5.7 nm by changing ALD cycle of the ZrO_2. On the other hand, the AZ stack structure was also prepared by the same ALD process. PDA of the ZA (ZA wPDA) and AZ stack structures (AZ wPDA) was carried out at 600°C for 1 min in N_2. Finally, a 50-nm-thick TE-TiN was deposited to fabricate MIM capacitor. On the other hand, PMA was finally performed at 600°C for 1 min in N_2 after fabricating the TiN/AZ/TiN (AZ wPMA) and TiN/ZA/TiN (ZA wPMA) capacitors fabricated without PDA. MIM capacitors with AZ (AZ w/o) and ZA (ZA w/o) without PDA and PMA were prepared as references.

The structure and morphology of the ZA and AZ stacks of MIM capacitor was observed using transmission electron microscope (TEM) and energy dispersive X-ray spectroscopy (EDS) using JEM-2100F. The crystal structure of ZrO_2 layer was analyzes by grazing-angle incidence X-ray diffraction (GI-XRD, Rigaku SmartLab). Capacitance-voltage $(C\text{-}V)$ and current-voltage $(I\text{-}V)$ measurements were carried out at room temperature under ambient conditions using as a semiconductor device analyzer (Agilent B1500A) and a semiconductor parameter analyzer (Keithley 4200-SCS), respectively. $C\text{-}V$ measurements were performed in the range of -0.5 to 0.5 V at a frequency of 100 kHz. The k-value of the ZrO_2 was calculated using capacitance at 0 V of the $C\text{-}V$ characteristics.

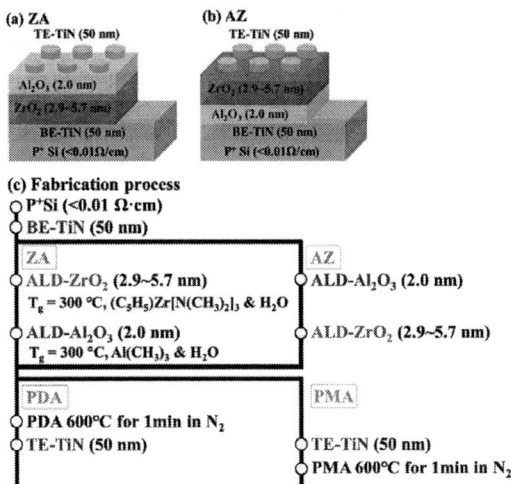

Figure 1. Schematic illustrations of (a) BE-TiN/ZA/TE-TiN and (b) BE-TiN/AZ/TE-TiN capacitors, and (c) fabrication process flow of MIM capacitors.

Results and Discussion

Figure 2(a) shows a cross-sectional TEM image and EDS elemental lines of Zr, Al and Ti atoms for ZA wPDA. The thicknesses of the ZA were 5.7/2.0 nm. The crystallization of ZrO_2 layer was observed while Al_2O_3 layer was kept o be an amorphous structure. The interface between the ZrO_2 and Al_2O_3 layers is found to be relatively separated without intermixing of Zr and Al atoms. The AZ w PDA was also observed to keep stack structure of an amorphous Al_2O_3 and a polycrystalline ZrO_2 as shown in **figure 2(b)**.

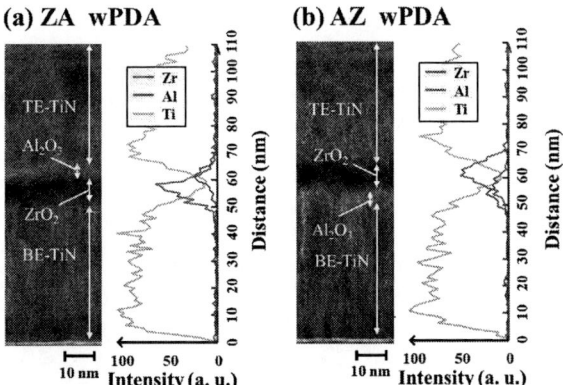

Figure 2. Cross-sectional TEM images and EDS elemental lines of Zr, Al and Ti atoms for (a) ZA wPDA and (b) AZ wPDA capacitors.

The structure of the ZA and AZ stacks were evaluated using XRD analysis after PDA and PMA. **Figure 3(a) and 3(b)** show the XRD patterns of ZA and AZ capacitors, respectively. One sharp peak at $2\theta = 36.4°$ was assigned to (111) plane of TiN. In the case of the w/o capacitors, no peak was observed in both of ZA and AZ, indicating that as-grown ZrO_2 layer had an amorphous structure. On the other hand, a broad peak appeared at $2\theta = 29.95 \sim 30.05°$, which was related to the orthorhombic (O), tetragonal (T) and cubic (C) phases, in the ZA and AZ stacks after PDA and PMA. No difference of the peak intensity and position were observed. Here, it is difficult to separate the peak at $2\theta = 29.95 \sim 30.05°$ into separate peaks for O, T, and C phases because the peak positions are extremely close. The structure of the ZrO_2 layer was found to change from an amorphous to a polycrystalline regardless of stack structure and annealing process.

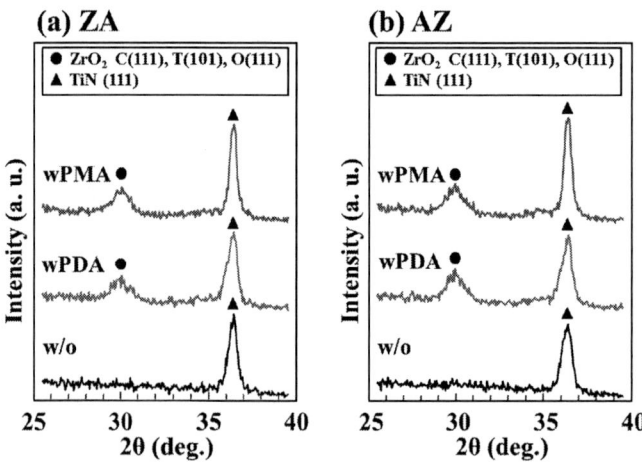

Figure 3. XRD patterns of (a) the ZA w/o, ZA wPDA and ZA wPMA, and (b) the AZ w/o, AZ wPDA and AZ wPMA capacitors.

Figure 4(a)-4(d) show the *C-V* characteristics of ZA and AZ capacitors. Voltage applied from -0.5 to 0.5 V at 100 kHz. All capacitors exhibited sufficiently small tan δ of below 0.2, indicating that ZA and AZ insulators have superior insulating property. The capacitance of the ZA wPMA, AZ wPMA and AZ wPDA capacitors reasonably decreased as the ZrO_2 layer thickness increased from 2.9 to 5.7 nm. On the other hand, surprisingly, the reduction of capacitance in the ZA wPDA capacitor was suppressed, especially, in the thicker ZrO_2 thickness region of above 4.9 nm.

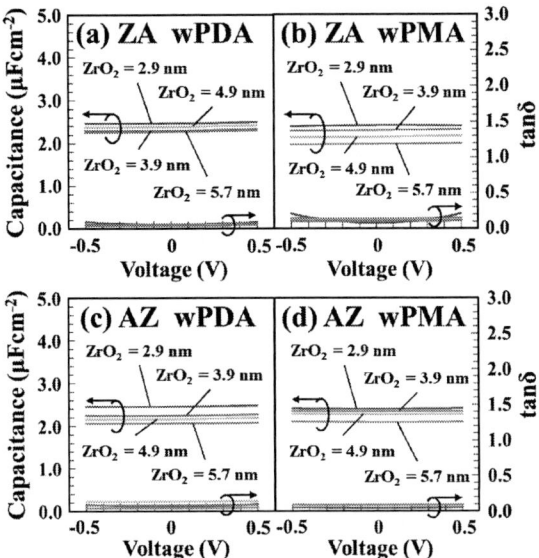

Figure 4. C-V characteristics of (a) ZA wPDA, (b) ZA wPMA, (c) AZ wPDA and (d) AZ wPMA capacitors. The thickness of the ZrO_2 layer was 2.9, 3.9, 4.9 and 5.7 nm.

Figure 5(a) shows a comparison of capacitance at 0 V for the ZA wPDA, ZA wPMA, AZ wPDA and AZ wPMA capacitors. The capacitance of the AZ wPMA and AZ wPDA decreased to 2.1 μFcm^{-2} as the ZrO_2 layer thickness increased from 2.9 to 5.7 nm. On the other hand, the ZA wPDA capacitor exhibited almost similar capacitance while the capacitance of the ZA wPMA capacitor decreased to 2.0 μFcm^{-2} as the ZrO_2 thickness increased. **Fig. 5(b)** shows a comparison of k-value of ZrO_2 layer for the ZA wPDA, ZA wPMA, AZ wPDA and AZ wPMA. The k-value of Al_2O_3 layer, estimated from capacitance of TiN/Al_2O_3/TiN capacitors which a 6-nm-thick Al_2O_3 layer was deposited by ALD 300°C, at 0V was found to be 7.7. The k-value of ZrO_2 layer was estimated from capacitance at 0V based on this k-value of Al_2O_3. The k-values of ZrO_2 layer of all w/o capacitors exhibited about 22-26 (not shown). The k-values of all capacitors gradually increased from 26-29 to 32-37 as the ZrO_2 thickness increased from 2.9 to 4.9 nm. W. Weinreich et al. reported that the k-value increased with increasing the thickness of the ZrO_2 for ZAZ structure (2). This is because the bulk-like characteristic of the ZrO_2 can be obtained. On the other hand, the k-value of the ZA wPDA dramatically jumped to 44 when the ZrO_2 thickness was 4.9 nm. This value was about 12 larger than those of the ZA wPMA with the same ZA structure. Furthermore, it was 11 larger than the AZ wPDA under the same PDA process.

Here, we consider the reason why the ZA wPDA can be obtained high k-value. We paid attention to different structures of BE-TiN/ZA and BE-TiN/ZA/TE-TiN under PDA and PMA, respectively. The coefficients of thermal expansion (CTE) of the Al_2O_3, ZrO_2 and TiN were 5.4~8.4 × 10^{-6}, 8.0~11.0 × 10^{-6}, and 9.4~10.3 × 10^{-6} K^{-1}, respectively (8-10). ZrO_2 had a similar CTE value of TiN and larger CTE value than Al_2O_3. For the ZA wPDA capacitor, ZrO_2 layer was annealed with an asymmetric structure sandwiched between

Al_2O_3 and BE-TiN with different CTE values during PDA. This asymmetric structure may introduce some stress of ZrO_2 layer and results in grain growth and crystallization of high-k phase. For the AZ wPDA, the ZrO_2 layer grew freely without stress and results in a small k-values below 34. In case of ZA wPMA and AZ w PMA, the ZrO_2 layer of AZ and ZA became symmetric structure by sandwiching between BE-TiN and TE-TiN during PMA and no high k-value was obtained like the ZA wPDA.

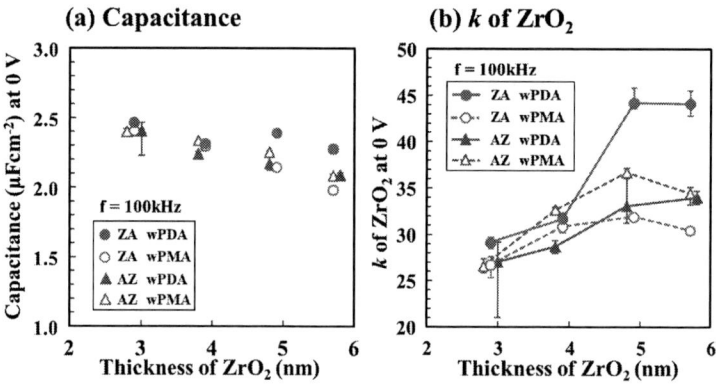

Fig. 5. (a) Capacitance at 0 V and (b) k-value of ZrO_2 layer as a function of the thickness of ZrO_2 layer for ZA wPDA, ZA wPMA, AZ wPDA and AZ wPMA capacitors.

Fig. 6 shows the J–electric field (E) properties of the ZA wPDA, ZA wPMA, AZ wPDA and AZ wPMA capacitors. The thickness of the ZrO_2 layer was 5.7 nm. The breakdown voltage exhibited similar values of 3.8-4.0 V (4.9-5.1 MVcm^{-1}) for all capacitors. Next, E-value of each capacitor, defined at a $J = 1 \times 10^{-5}$ Acm^{-2}, was compared. The E-value increased in the following order: ZA wPMA = AZ wPMA (1.1-1.2 MVcm^{-1}) < AZ wPDA (1.4) < ZA wPDA (1.9). This indicated that the PDA treatment could produce insulator which have a low J than the PMA treatment.

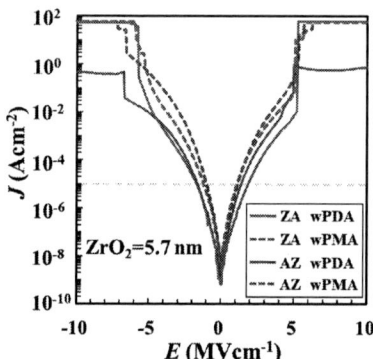

Fig. 6. J–E properties of the ZA wPDA, ZA wPMA, AZ wPDA and AZ wPMA capacitors. The thickness of the ZrO_2 layer was 5.7 nm.

Fig. 7 shows the relationship between the k-value of ZrO_2 and the E-value of each capacitor, defined at a $J = 1 \times 10^{-5}$ Acm^{-2}. The other capacitors except of the ZA wPDA with ZrO_2 layers (4.9-5.7 nm) showed a reasonable behavior of a slight increase of k-value and a remarkable increase of E-value as the ZrO_2 thickness increased. The ZA wPDA with ZrO_2 layers (4.9-5.7 nm) exhibited superior characteristics satisfying a high k-value and a high E-value.

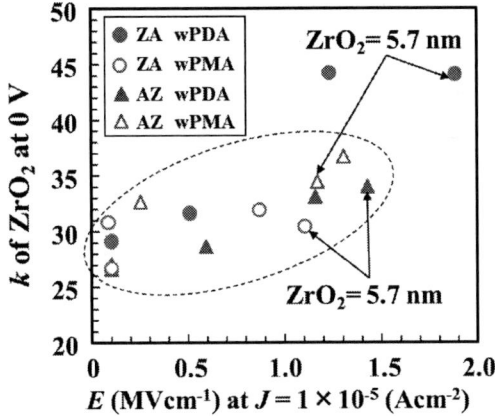

Fig. 7. Relationship between the k-value of ZrO_2 and the E-value of each capacitor, defined at a $J = 1 \times 10^{-5}$ Acm^{-2} for all capacitors.

Conclusion

We studied characteristics of the BE-TiN/ZA/TE-TiN and BE-TiN/AZ/TE-TiN capacitors after PDA and PMA at 600°C in N_2. The ZrO_2 layer with 2.9-5.7 nm thick had a polycrystalline structure and a 2-nm-thick Al_2O_3 layer had an amorphous structure. The k-value of the ZrO_2 layer in the ZA wPDA dramatically increased from 32 to 44 when ZrO_2 thickness changed from 3.9 to 4.9 nm while other capacitors exhibited saturated k-value around 32-37. This is due to the asymmetric structure which ZrO_2 layer sandwiched between BE-TiN and Al_2O_3 or symmetric structure which ZrO_2 layer sandwiched between BE-TiN and TE-TiN in the annealing of the ZrO_2 layer. The ZA wPDA with ZrO_2 layers (4.9-5.7 nm) showed superior characteristics satisfying a high k-value (44) and a high E-value (1.2-1.9 MVcm^{-1}) at $J = 1 \times 10^{-5}$ Acm^{-2}. Considering to ZAZ insulator fabrication of DRAM capacitor, PDA treatment should perform after 1st-ZrO_2/Al_2O_3 layer deposition to fabricate a high dielectric constant ZrO_2.

Acknowledgments

This work was partially supported by JSPS KAKENHI (Nos. JP20H02189 and JP18J22998). The authors thank all staff members of the Nanofabrication Group of NIMS for their support their support in the fabrication of the TiN/ZA/TiN capacitors.

References

1. Y.- H. Wu, C.-K. Kao, B.-Y. Chen, Y.-S. Lin, M.-Y. Li, and H.-C. Wu, *Appl. Phys. Lett.* **93**, 033511 (2008).
2. W. Weinreich, A. Shariq, K. Seidel, J. Sundqvist, A. Paskaleva, M. Lemberger, and A. J. Bauer, *J. Vac. Sci. Technol. B* **31**, 01A109 (2013).
3. T. Sawada, T. Nabatame, T. D. Dao, I. Yamamoto, K. Kurishima, T. Onaya, A. Ohi, K. Ito, M. Takahashi, K. Kohama, T. Ohishi, A. Ogura, and T. Nagao, *J. Vac. Sci. Technol. A* **35**, 061503 (2017).
4. T. Onaya, T. Nabatame, T. Sawada, K. Kurishima, N. Sawamoto, A. Ohi, T. Chikyow, and A. Ogura, *Thin Solid Films* **655,** 48 (2018).
5. D. Martin, M. Grube, W. Weinreich, J. Muller, W. M. Weber, U. Schroder, H. Riechert, and T. Mikolajick, *J. Appl. Phys.* **113**, 194103 (2013).
6. D. Martin, M. Grube, W. Weinreich, J. Muller, L. Wilde, E. Erben, W. M. Weber, J. Heitmann, U. Schroder, T. Mikolajick, and H. Riechert, *J. Vac. Sci. Technol. B* **29**, 01AC02 (2011).
7. S. Y. Lee, J. Chang, Y. Kim, H. J. Lim, H. Jeon, and H. Seo, *Appl. Phys. Lett.* **105**, 201603 (2014).
8. H. Holleck, *J. Vac. Sci. Technol. A* **4**, 2661 (1986).
9. P. Bouvier, E. Djurado, C. Ritter, A. J. Dianoux, and G. Lucazeau, *Int. J. Inorg. Mater.* **3**, 647 (2001).
10. Z. H. Cen, B. X. Xu, J. F. Hu, R. Ji, Y. T. Toh, K. D. Ye, and Y. F. Hu, *J. Phys. D: Appl. Phys.* **50**, 075105 (2017).

Study of SiO₂ Interfacial Layer Growth during Fabrication Process of Ferroelectric Hf$_x$Zr$_{1-x}$O$_2$-Based Metal-Ferroelectric-Semiconductor

T. Onaya[a, b, c], T. Nabatame[b], M. Inoue[b], T. Sawada[b], H. Ota[a], and Y. Morita[a]

[a] National Institute of Advanced Industrial Science and Technology (AIST),
1-1-1 Umezono, Tsukuba, Ibaraki 305-8568, Japan
[b] National Institute for Materials Science (NIMS),
1-1 Namiki, Tsukuba, Ibaraki 305-0044, Japan
[c] Research Fellow of Japan Society for the Promotion of Science (JSPS) PD,
5-3-1 Kojimachi, Chiyoda-ku, Tokyo 102-0083, Japan

E-mail: takashi.onaya@aist.go.jp

We studied a SiO₂ interfacial layer (SiO₂-IL) growth between a ferroelectric Hf$_x$Zr$_{1-x}$O$_2$ (HZO) film, which fabricated by atomic layer deposition (ALD) using a Hf/Zr cocktail precursor, and p-type Si substrate during the metal-ferroelectric-semiconductor (MFS) fabrication. For the annealing process, two types of a post-deposition annealing (PDA) and post-metallization annealing (PMA) were employed before and after the fabrication of the TiN top-electrodes, respectively. An ultra-thin SiO₂-IL of one or two monolayers at the HZO/Si interface was achieved using a 300°C fabrication process of ALD, and PDA or PMA. In contrast, PDA and PMA processes at ≥ 400°C led to an unexpected SiO₂-IL formation, which increased with the annealing temperature, regardless of the formation of TiN top-electrodes. Therefore, we found that a suppressed SiO₂-IL of HZO-based MFS structures can be obtained using the 300°C fabrication process.

Introduction

Recently, ferroelectric Hf$_x$Zr$_{1-x}$O$_2$ (HZO) thin films have attracted much attention towards high-density and low-power ferroelectric memories of ferroelectric field-effect transistors (FeFETs) as synapse devices for neuromorphic systems. (1–3) An annealing process at ≥ 300°C is typically required to form the ferroelectric orthorhombic (O) phase of HZO films. (4–6) It remains a big issue of an abnormal SiO₂ interfacial layer (SiO₂-IL) growth at the HZO/Si interface during a metal-ferroelectric-semiconductor (MFS) fabrication process such as ferroelectric film deposition and annealing processes. It has been reported that the increase of a SiO₂-IL thickness caused the remanent polarization (P_r) reduction of the MFS capacitors because the dielectric constant (k = 3.9) of SiO₂ is much lower than that (k = 20–40) of HZO. (7, 8) Moreover, a SiO₂-IL growth led to degradation of reliability in MFS capacitors. However, it is unclear how the process conditions of MFS structures, such as an atomic layer deposition (ALD) process of HZO films and annealing process, affect a SiO₂-IL formation. In this work, we systematically investigated a SiO₂-IL growth as a function of the process temperature of ALD and annealing processes in TiN/HZO/SiO$_2$-IL/p-Si MFS structures using X-ray photoelectron spectroscopy (XPS) and transmission electron microscopy (TEM) analyses.

Experimental

The fabrication process flow of the ALD-HZO/SiO$_2$-IL/p-Si samples for XPS analysis is shown in **figure 1(a)**. First, p-type Si substrate was cleaned using buffered hydrofluoric acid (BHF) for 30 sec to remove native oxide, and then a 3-nm-thick HZO film was deposited by ALD using (Hf/Zr)[N(C$_2$H$_5$)CH$_3$]$_4$ (Hf:Zr = 1:1) cocktail precursor and H$_2$O gases. The ALD growth temperature of HZO films was varied from 200 to 300°C. Two types of annealing processes such as a post-deposition annealing (PDA) and post-metallization annealing (PMA) were carried out at 300–600°C in an N$_2$ atmosphere before and after the deposition of a 10-nm-thick TiN film by DC sputtering, respectively, as shown in **figure 1(b)** and **1(c)**. For the PMA-treated samples, a TiN film was removed after the PMA process for XPS analysis. A TiN/HZO (10 nm)/SiO$_2$-IL/p-Si MFS structure was also fabricated using the 300°C fabrication process of ALD and PMA for TEM analysis.

A SiO$_2$-IL growth was evaluated by XPS Si 2p measurements using an Al-Kα monochromatic source. XPS spectra were referenced to the Si 2p$_{3/2}$ peak at a binding energy of 99.4 eV. In addition, the intensity of XPS spectra was normalized using the Si 2p$_{3/2}$ peak. The thickness and morphology of the HZO films were evaluated using spectroscopic ellipsometry and cross-sectional TEM images.

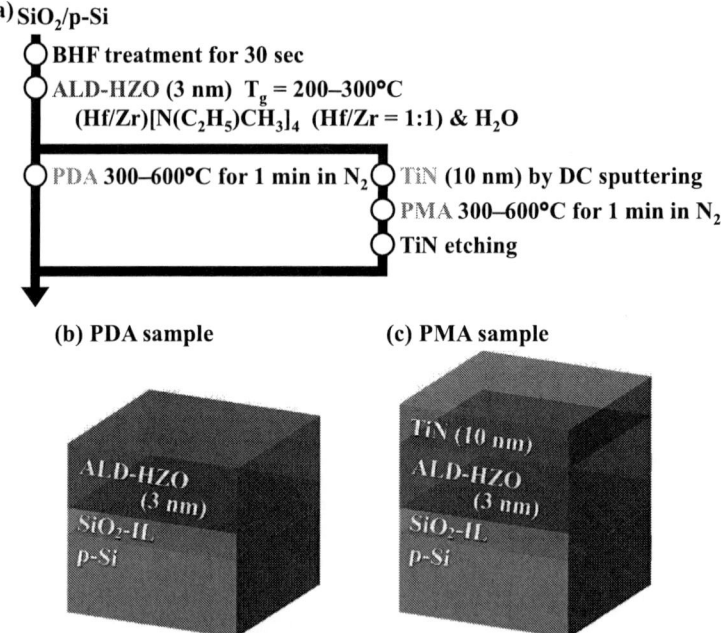

Figure 1. (a) Fabrication process flow and schematic illustration of (b) the PDA- and (c) PMA-treated samples for XPS analysis.

Results and Discussion

Figure 2 shows the XPS spectra of Si 2p core-level for the ALD-HZO (3 nm)/Si samples prepared with different ALD growth temperatures from 200 to 300°C. The XPS Si 2p measurement of a BHF-treated Si substrate was also carried out as a reference. Two noticeable peaks attributed to the Si $2p_{3/2}$ and Si $2p_{1/2}$ were observed at 99.4 and 100.0 eV, respectively. All of the XPS Si 2p spectra were similar regardless of the ALD growth temperatures, which well matched that of a BHF-treated Si substrate. Moreover, no significant peak originating from the Si–O component was observed in all XPS spectra of Si 2p core-level. Therefore, we found that the formation of a SiO_2-IL at the ALD-HZO/p-Si interface was negligible small in the temperature range of the ALD process at 200–300°C. Based on these data, we employed 300°C as an ALD growth temperature of HZO films for the fabrication of MFS structures.

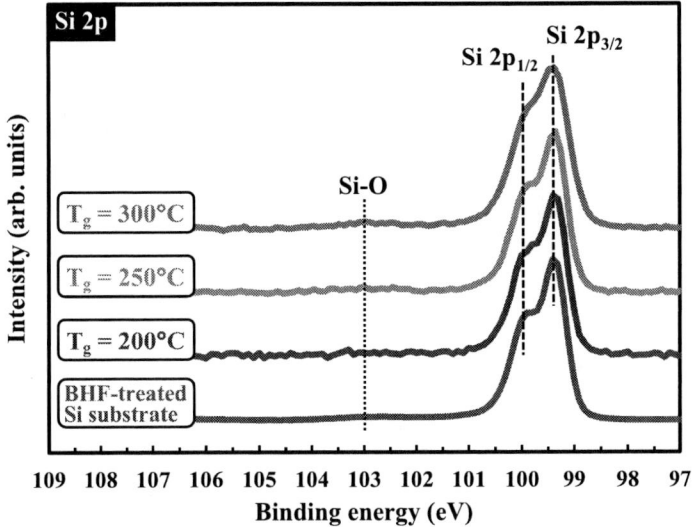

Figure 2. XPS spectra of Si 2p core-level for the ALD-HZO (3 nm)/Si samples prepared with different ALD growth temperatures from 200 to 300°C.

Figure 3 shows the XPS spectra of Si 2p core-level for the ALD-HZO (3 nm)/Si samples prepared with different PDA temperatures from 300 to 600°C. An HZO film was deposited on a BHF-treated Si substrate by ALD at 300°C. The peak originating from Si–O bonding was clearly observed at approximately 103.0 eV, which corresponds well with the previous reports, when the PDA temperature increased over 400°C. (9) In addition, the intensity of the Si–O peak increased with the PDA temperature, indicating that the thickness of a SiO_2-IL between the HZO film and p-Si substrate increased. On the other hand, the sample after PDA process at 300°C showed almost the same XPS Si 2p spectrum as the as-grown sample. Thus, we concluded that a suppressed SiO_2-IL at the ALD-HZO/p-Si interface was achieved using the 300°C-PDA process.

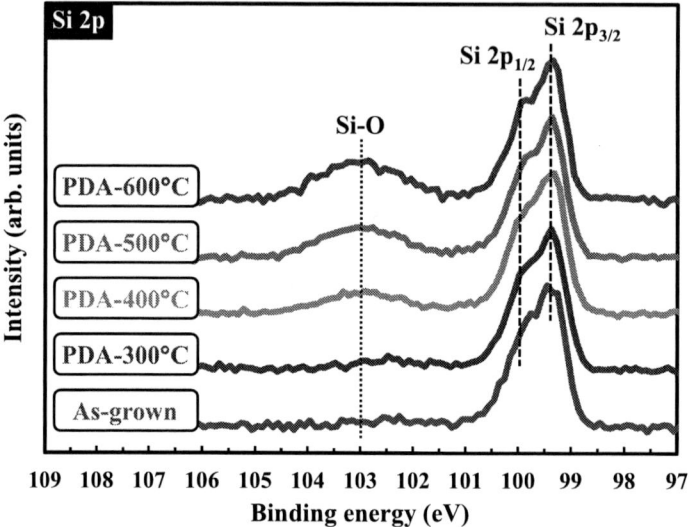

Figure 3. XPS spectra of Si 2p core-level for the ALD-HZO (3 nm)/Si samples prepared with different PDA temperatures from 300 to 600°C.

Figure 4 shows the XPS spectra of Si 2p core-level for the ALD-HZO (3 nm)/Si samples prepared with different PMA temperatures from 300 to 600°C. An HZO film was deposited on a BHF-treated Si substrate by ALD at 300°C. For the PMA samples, the TiN film was first deposited on the HZO film, and then the TiN film was removed after the PMA process. One broad peak that corresponds to the Si–O component was also observed at 103.0 eV when the PMA temperature was 400°C, while that of 300°C-PMA-treated sample was negligible small. In addition, the peak area of the PMA-treated samples increased with the PMA temperature, which was comparable with that of the PDA-treated samples when subjected to the same annealing temperature. Therefore, a suppressed SiO_2-IL growth was achieved using the PMA process at 300°C, which is the same amount as that of the as-grown sample. Moreover, these results suggested that a SiO_2-IL growth was dominantly affected by the process temperature. It has been reported that TiN electrodes provide tensile stress to the HZO film during the PMA process, promoting the formation of the metastable ferroelectric O phase. (10) Thus, the PMA process was employed to form the HZO-based MFS structures instead of the PDA process.

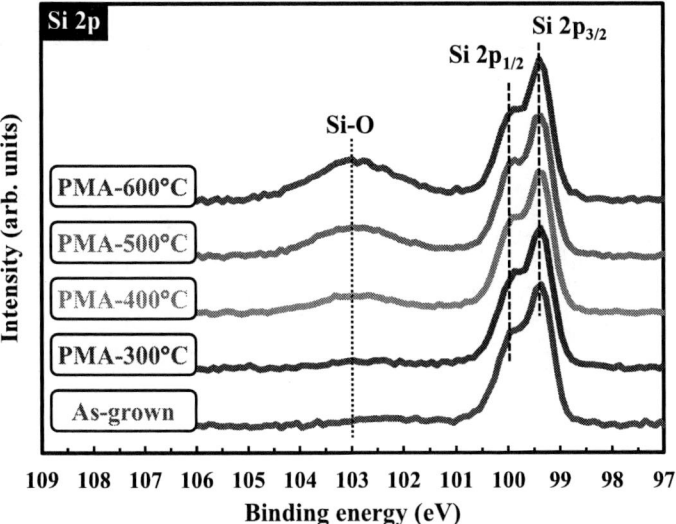

Figure 4. XPS spectra of Si 2p core-level for the ALD-HZO (3 nm)/Si samples prepared with different PMA temperatures from 300 to 600°C. The TiN film was first deposited on the HZO film, and then the TiN film was removed after the PMA process.

Figures 5(a) and **5(b)** show the cross-sectional TEM images of the as-grown and 300°C-PMA-treated MFS structures, respectively. A 10-nm-thick HZO film was deposited on a BHF-treated Si substrate as a ferroelectric film of the MFS structure by ALD at 300°C. An ALD-HZO film with the thickness of ~10 nm remained conformal and uniform between the p-Si substrate and TiN top-electrode even after the PMA process. The HZO film of the as-grown and 300°C-PMA-treated MFS structures partially formed nanocrystals with a grain size of 5–10 nm, while most of the film kept an amorphous structure. For the as-grown MFS structure, an ultrathin SiO_2-IL with the thickness of approximately 0.5 nm was found to be formed at the ALD-HZO/p-Si interface, which is equivalent to one or two monolayers of the SiO_2 film. Furthermore, the MFS structure after the PMA process at 300°C maintained the same SiO_2-IL thickness as that of the as-grown case, which corresponds well with the results of the XPS analysis as shown in **figure 4**. Therefore, we concluded that the annealing process at a low temperature of 300°C suppressed a SiO_2-IL growth for the fabrication of HZO-based MFS structures.

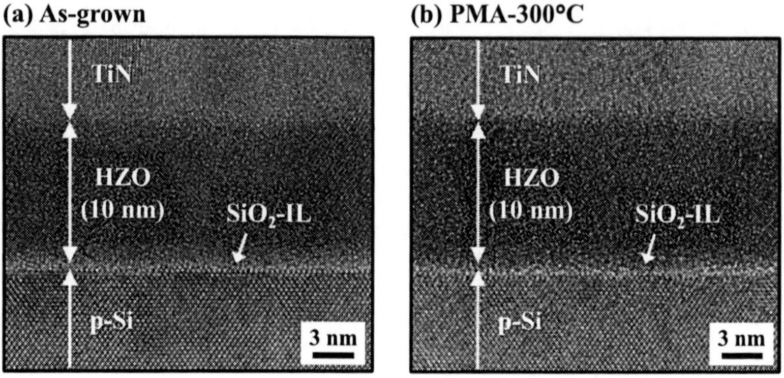

(a) As-grown

(b) PMA-300°C

Figure 5. Cross-sectional TEM images of (a) the as-grown and (b) 300°C-PMA-treated MFS structures, respectively, with a 10-nm-thick ALD-HZO film.

Conclusion

The formation of a SiO_2-IL at the ALD-HZO/p-Si interface during the ALD and annealing processes was systematically investigated. No noticeable peak attributed to the Si–O bonding for the ALD-HZO/SiO_2-IL/p-Si samples was observed regardless of the ALD growth temperature from 200 to 300°C evaluated by XPS analysis, indicating that the formation of a SiO_2-IL was negligible small during the ALD process. For the PDA- and PMA-treated samples, 300°C-annealed samples showed almost the same XPS spectra as that of the as-grown sample, which thickness was equivalent to one or two monolayers determined by cross-sectional TEM images. In contrast, the Si–O peak of the both PDA- and PMA-treated samples increased when the annealing temperature was above 400°C. Thus, low temperature PDA and PMA processes at 300°C could prevent a SiO_2-IL growth regardless of the fabrication of TiN top-electrodes. Based on these experimental results, the 300°C fabrication process of HZO-based MFS structures is one of the promising candidates for future ferroelectric nonvolatile memory applications.

Acknowledgments

This work was partially supported by JSPS KAKENHI (Grant Nos. JP21J01667 and JP20H02189). The authors thank all staff members of the Nanofabrication Group of NIMS for their support in fabricating the MFS structures.

References

1. J. Y. Kim, M.-J. Choi, and H. W. Jang, *APL Mater.* **9**, 021102 (2021).
2. M.-K. Kim, I.-J. Kim, and J.-S. Lee, *Appl. Phys. Lett.* **118**, 032902 (2021).
3. S. Oh, H. Hwang, and I. K. Yoo, *APL Mater.* **7**, 091109 (2019).
4. T. Onaya, T. Nabatame, Y. C. Jung, H. Hernandez-Arriaga, J. Mohan, H. S. Kim, N. Sawamoto, C.-Y. Nam, E. H. R. Tsai, T. Nagata, J. Kim, and A. Ogura, *APL Mater.* **9**, 031111 (2021).
5. T. Onaya, T. Nabatame, N. Sawamoto, A. Ohi, N. Ikeda, T. Nagata, and A. Ogura, *Microelectron. Eng.* **215**, 111013 (2019).
6. H. J. Kim, Y. An, Y. C. Jung, J. Mohan, J. G. Yoo, Y. I. Kim, H. Hernandez-Arriaga, H. S. Kim, J. Kim, and S. J. Kim, *Phys. Status Solidi RRL* **15**, 2100028 (2021).
7. J. Mohan, H. Hernandez-Arriaga, Y. C. Jung, T. Onaya, C.-Y. Nam, E. H. R. Tsai, S. J. Kim, and J. Kim, *Appl. Phys. Lett.* **118**, 102903 (2021).
8. K. Toprasertpong, K. Tahara, T. Fukui, Z. Lin, K. Watanabe, M. Takenaka, and S. Takagi, *IEEE Electron Device Lett.* **41**(10), 1588 (2020).
9. E. Maeda, T. Nabatame, M. Hirose, M. Inoue, A. Ohi, N. Ikeda, and H. Kiyono, *J. Vac. Sci. Technol. A* **38**, 032409 (2020).
10. S. J. Kim, D. Narayan, J.-G. Lee, J. Mohan, J. S. Lee, J. Lee, H. S. Kim, Y.-C. Byun, A. T. Lucero, C. D. Young, S. R. Summerfelt, T. San, L. Colombo, and J. Kim, *Appl. Phys. Lett.* **111**, 242901 (2017).

Chapter 6

Novel Process-Growth

Cutting-edge epitaxial processes for sub 3 nm technology nodes: application to nanosheet stacks and epitaxial wrap-around contacts

A. Hikavyy, C. Porret, R. Loo, M. Mencarelli, P. Favia, M. Ayyad, B. Briggs, R. Langer and N. Horiguchi

Imec, Kapeldreef 75 Leuven Belgium: Andriy.Hikavyy@imec.be

> This work reports on low temperature epitaxial growth solutions for the processing of advanced CMOS devices beyond the 3 nm technological node. The complex stacking of highly compositionally contrasted strained group IV materials is first demonstrated at 500°C. It enables the formation of active nanosheet channels with bottom isolation, necessary for ultimate transistor scaling. Using high order Si and Ge precursors also offers great opportunities for the epitaxy of advanced source/drain materials. It allows achieving hole active concentrations as high as 1.3×10^{21} cm^{-3} in in-situ B-doped $Si_{0.5}Ge_{0.5}$, providing Ti / SiGe:B contacts with low specific resistivity. Grown at temperatures as low as 400°C, the epilayers deposit in a conformal manner, thereby wrapping high aspect ratio 3-dimensional structures and maximizing the contact areas, being an additional option to further decrease device access resistances. As an alternative to B-doping only, we also demonstrate *uniformly* Ga-doped materials with concentrations ~ 1×10^{20} cm^{-3}. B and Ga are finally combined to co-dope SiGe and further reduce the Ti / p-SiGe contact resistivity.

Introduction

The epitaxial growth of group IV semiconductor materials is one of the backbones of modern IC production flows. At present nodes it is used for the Source/Drain (S/D) engineering of both p- and n-MOS high performance transistors, allowing the deposition of SiGe:B and Si:P layers with active carrier concentrations ~ 1×10^{21} cm^{-3}, ensuring low contact resistivities [1]. However, with reducing device dimensions, the growing importance of contact resistance in devices remains a major concern. A significantly lower parasitic S/D resistance could also be obtained thanks to the use of *epitaxial* and metal wrapped-around contacts (WAC) [2]. In addition, alternatives to Si channel materials (SiGe or Ge) and new device concepts such as 3D transistor stacking, complementary FET (CFET) and gate-all-around (GAA) nanosheet devices, all considered for the upcoming technological nodes of 3 nm and below [3,4], require process temperatures ≤ 500°C.

To meet these temperature requirements, epitaxial processes based on the combination of high order Si and Ge precursors are explored together with low temperature Cl_2 etching [5,6]. The strength of this approach has already been demonstrated by the successful implementation of Ge/SiGe stacks and Ge S/D deposited at low temperature for the production of pMOS Ge GAA devices exhibiting excellent electrostatic control at sub-30 nm gate lengths [3].

In this contribution, we use advanced low temperature epitaxial processes to enable cutting-edge applications. Complex and highly strained multi stacks are pseudomorphically grown with a limited thermal budget. The compositional contrast between the different parts of the stack, combined with very sharp interfaces, allows for the selective removal of sacrificial layers and the formation of active channels being totally isolated from parasitic channels in the substrate. In addition, highly B, Ga or B+Ga co-doped SiGe layers are grown at temperatures down to 400°C with an active doping concentration exceeding 1×10^{21} cm^{-3}. Finally, epitaxial WAC structures are demonstrated.

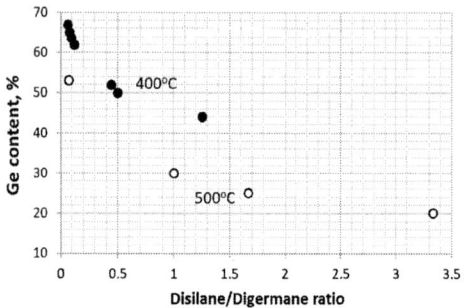

Figure 1. Ge content as function of precursors flow ratio for SiGe grown with disilane and digermane at 400 and 500°C.

Results and discussion

A. Epitaxy of highly strained and contrasted stacks for nanosheet devices with bottom isolation

Lateral gate-all-around (GAA) MOSFETs based on nanowires and/or nanosheets (NW/NSH) are promising candidates for future CMOS technology nodes due to enabling aggressive gate pitch scaling because of better electrostatic control [7]. Furthermore, a partial and full bottom dielectric isolation have been proposed for nanosheet devices to improve device performance [8].

A first challenge in making nanosheet devices with bottom isolation is to avoid plastic relaxation caused by the cumulative strain in the $Si/Si_{0.5}Ge_{0.5}/2x[Si_{0.8}Ge_{0.2}/Si]$ epi stacks targeted in this work. Low temperature epitaxial growth and acceptable growth rates are required to minimize the total thermal budget. This is achieved by using Si_2H_6 and Ge_2H_6 at temperatures $\leq 500°C$ [5]. Growth rates ≥ 5 nm/min and ~ 0.5 nm/min are obtained, respectively, for the SiGe and Si layers constituting the stack with a wide range of possible Ge concentrations (**Fig. 1**).

Fig. 2 summarizes cross-section TEM and energy dispersive X-ray spectroscopy (EDS) data from the grown stack. The cross-section TEM displayed in **Fig. 2a** confirms the high structural quality of the different layers and the presence of very sharp interfaces between them. HR-XRD and AFM measurement results (not shown here) support the absence of strain relaxation. The EDS linescan presented in **Fig. 2b** provides quantitative information about the composition of the different layers and the sharpness of interfaces between them. It confirms Ge concentrations of ~ 50 and 20% for the bottom

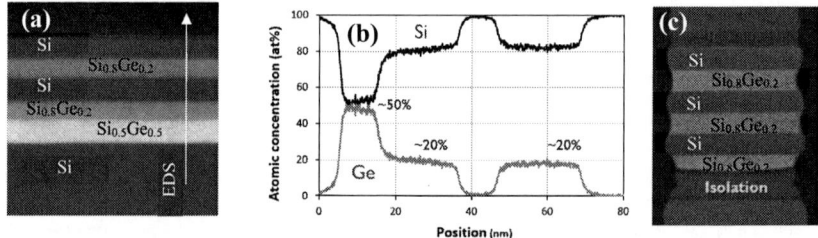

Figure 2. (a) HAADF-STEM cross-sectional image and (b) corresponding EDS linescan, acquired along the arrow indicated in Fig. 3a, of a $Si/Si_{0.5}Ge_{0.5}/2x[Si_{0.8}Ge_{0.2}/Si]$ stack grown at 500°C. (c) HAADF-STEM image of the final nanosheet device structure with integrated bottom SiN/SiCO isolation.

and top SiGe layers, matching the targeted values. Following this first epi step, the bottom layer is selectively removed after fin patterning by an ammonia-peroxide mixture with ~10:1 etch selectivity with respect to $Si_{0.8}Ge_{0.2}$. A careful optimization of the selective etch processes and of the Ge content in the epitaxial layers is required to reduce the SiGe loss of the sacrificial layers used for final channel release. The remaining cavity filled with conformally deposited SiN/SiCO (**Fig. 2c**). This allows to suppress a parasitic channel under the lowest wire/sheet channel and a possibility of back biasing [9].

The proposed layer replacement scheme has been finally demonstrated by integration of forksheet n- and pFETs co-integrated with gate all around nanosheet FETs reported in [10].

B. In-situ doped SiGe grown at 400°C with improved electrical material properties

Adaptation of complex $Si/Si_{0.5}Ge_{0.5}/2x[Si_{0.8}Ge_{0.2}/Si]$ stacks presented above by device integration sets stringent requirements to all consequent processing steps. For the Source/Drain epitaxy it means removal of any pre-epi bakes, which are typically done at temperatures around 800°C, and limits process temperature to maximum 500°C in order to avoid Si/SiGe stack relaxation. Use of conventional precursors and etchants in this temperature range becomes challenging since growth rates are reduced, promoting switching to more exotic high order Si and Ge precursors and even metal-organic precursors, which are typically avoided due to risk of C incorporation in the final epi layers.

Growing SiGe at reduced temperature (400°C) allows to broaden the process window for in-situ boron doping and yields a ~ 30% higher active carrier concentration (1.3×10^{21} cm^{-3}) in comparison to conventional SiGe:B epi schemes [11]. However, the process is found to add a next level of complexity, namely a very strong competition between Ge and B incorporation (**Fig. 3**), unseen for conventional precursors. This effect is effectively circumvented by finetuning the disilane/digermane flow ratio to obtain the required Ge concentration.

Circular transmission line measurements (CTLM) on blanket layers show that the contact resistivity of low temperature SiGe:B significantly outperforms our high temperature blanket SiGe:B (B ~ 4×10^{20} cm^{-3}) process of record (POR) based on conventional precursors (**Fig. 4**). The reported contact resistivities have been obtained on as-grown layers without post-epi implants nor post-epi thermal treatments.

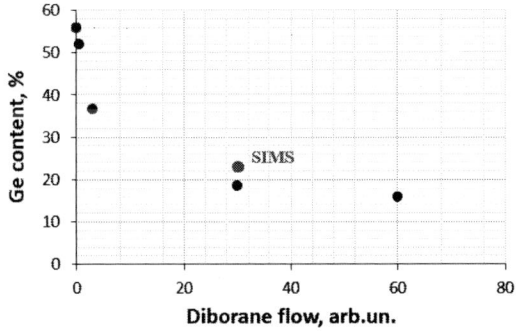

Figure 3. Ge apparent concentration extracted by HRXRD as function of diborane flow for SiGe:B layers grown at 400°C, showing a strong competition between the incorporation of Ge and B. An additional SIMS measurement shows that strain compensation by high B concentration is small compared to the competition effect.

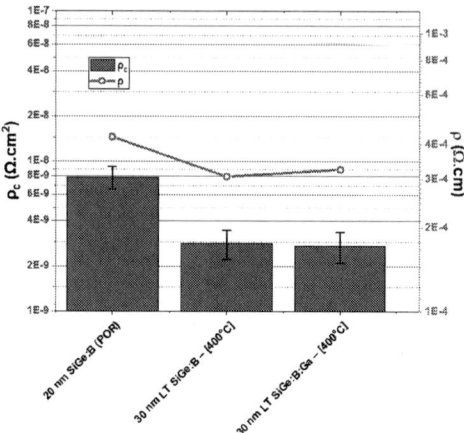

Figure 4. Resistivity (ρ) and contact resistivity (ρ_c) comparison between POR SiGe:B and low temperature SiGe:B, SiGe:B,Ga.

SiGe co-doping with B and Ga showed sub-10^{-9} Ohm.cm^2 contact resistivities [12,13]. Nevertheless, very little information on in-situ doping with Ga is available in literature. Growing a Ga-doped SiGe layer with high active concentration using the standard epitaxy processes is challenging due to very small equilibrium segregation coefficients of Ga in both Si and Ge causing a strong Ga segregation toward the surface [13]. Low temperature SiGe:Ga growth dramatically improves the situation resulting in layers with a high active Ga concentration up to ~1×10^{20} cm^{-3} and uniform Ga distribution through the complete $Si_{0.4}Ge_{0.6}$ layer (**Fig. 5**). SiGe co-doping with B and Ga with high dopants concentrations is further demonstrated in **Fig. 6**. This result opens opportunities for setting up the selective deposition of SiGe:B:Ga following a cyclic deposition-etch routine, which is not possible when Ga segregates to

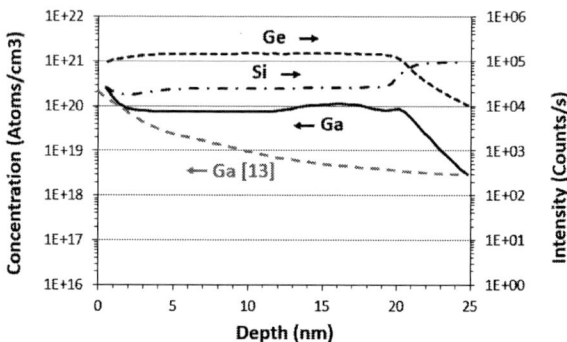

Figure 5. Ge, Si and Ga concentration profiles extracted by SIMS measurements performed on a SiGe:Ga layer grown at low temperature, resulting in a carrier concentration of ~1×10^{20} cm^{-3} and a uniform Ga distribution through the complete SiGe layer. The curve extracted from reference [13] shows a graded Ga profile, typically reported in literature.

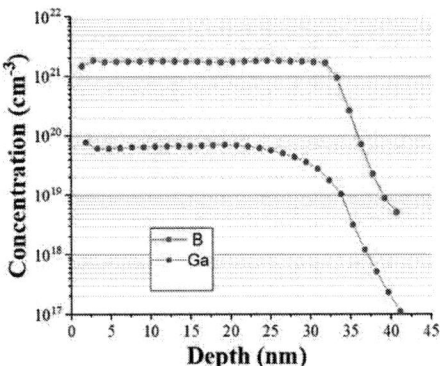

Figure 6. B and Ga concentration profiles extracted by SIMS from a $Si_{0.56}Ge_{0.44}$ epi layer co-doped with B and Ga. Uniform distribution of Ga allows setting up a selective cyclic deposition and etch SiGe:B,Ga process, which is not possible when Ga segregates to the surface during deposition.

the surface as in [13], as Ga dopants are consumed during the etching steps. The contact resistivity as extracted for the low temperature Ga+B co-doped SiGe process is similar to the value obtained for low temperature SiGe:B and significantly lower than the high temperature SiGe:B POR which is based on conventional precursors (**Fig. 4**). Furthermore, if necessary, SiGe:B:Ga, epitaxial growth can be combined with a laser annealing to promote movement of the accumulated during epitaxy Ga, towards the surface [14].

C. Wrap around S/D contacts enabled by low temperature selective epitaxial SiGe:B growth

Selectivity of epitaxial layers is typically achieved by adding HCl to the growth chemistry, resulting in etching materials deposited on oxide and nitride surfaces. Since HCl becomes completely ineffective, Cl_2 is used as an etching gas for selective epi schemes below 450°C [6, 15] and the co-flow approach is replaced by a cyclic deposition/etch approach.

Conventional selective S/D processes manifest a strong faceting along (111) Si planes, defining the well known "diamond" S/D shape (**Fig. 7**) found in existing fin FET transistors. The newly developed low temperature cyclic selective SiGe:B process described previously in [16] allows to control the faceting and obtain similar layer thicknesses for different surface orientation. This leads to

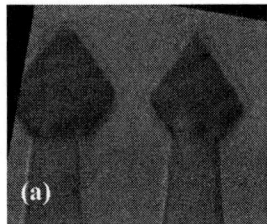 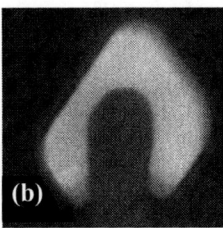 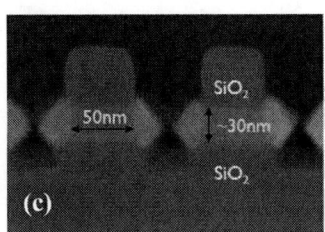

Figure 7. Typical "diamond" S/D shape formed after conventional epi growth presented for different cases: (a) embedded (finFET), (b) raised (finFET), (c) nano sheet FET (test structure).

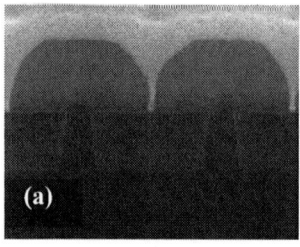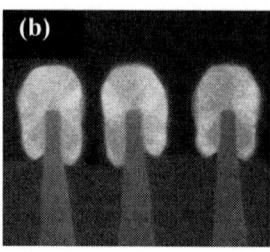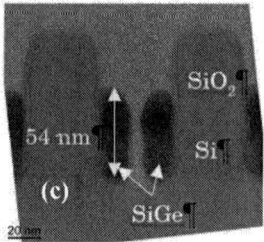

Figure 8. Novel S/D shape developed after low temperature cyclic SiGe:B growth at 400°C presented for different cases: (a) embedded (finFET), (b) raised (finFET), (c) nano sheet FET (test structure).

S/D layers with completely different and novel shapes (**Fig. 8**). The usual faceting is suppressed, resulting in the conformal deposition of the S/D material. The obtained S/D shape has a great potential for both finFET (**Fig. 8 a,b**) and nano/fork sheet (**Fig. 8c**) structures since it allows to eliminate voids due to merging between different S/D areas of the transistors standing next to each other and leads to the formation of in-situ doped wrap-around contacts.

The latter results in an increased contact area which reduces parasitic resistance (**Fig. 9**) [2]. Growth studies on fin structures with different surface orientations provide insight in the growth kinetics. This allows to quantify the SiGe:B growth rate on different crystallographic planes and to demonstrate the differences in growth characteristics for conventional (POR) SiGe:B and low temperature cyclic SiGe:B processes (**Fig. 10**). For the reference process, using conventional precursors, the growth rate shows a well-known and significant dependency on surface orientation with a 32% lower growth rate on {110} Si compared to the growth rate on {100} Si surfaces. The dependence of the growth rate on surface orientation leads to the strong faceting seen in **Fig. 7**. Our new low temperature cyclic SiGe:B process shows a much smaller variation in growth rate for these two planes (**Fig. 10**). This explains the delayed faceting and the conformal surface morphology reported in **Fig. 8**.

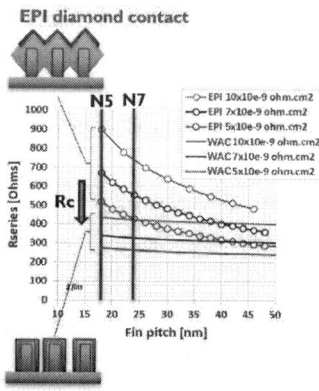

Figure 9. A comparison of contact resistance for "diamond" shaped and wrap around S/D contacts at scaled fin pitch. Wrap around contacts without S/D merging maintain large contact area at scaled fin pitch. [2].

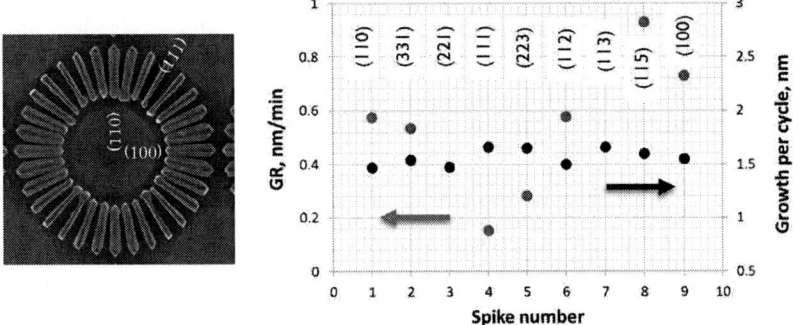

Figure 10. Fin structures with different orientations fabricated on (110) Si wafer allowing study growth rates on different Si planes (left). Variation in growth rates for a conventional (POR) SiGe:B (blue circles), and a low temperature SiGe:B cyclic processes (black circles).

Conclusions

We introduce a new (selective) epitaxial growth scheme for undoped and doped SiGe with modified growth characteristics. High order Si and Ge precursors and extremely low growth temperatures ($\leq$ 500°C) are used to deposit the active layers of nanosheet devices and their S/D. The deposition of high quality $Si/Si_{0.5}Ge_{0.5}/2x[Si_{0.8}Ge_{0.2}/Si]$ stacks enables the fabrication of high mobility nanosheet channels with bottom isolation. We demonstrate the low temperature epitaxy of various p-SiGe materials, exhibiting increased carrier concentrations with respect to conventional processes, and offering opportunities for the formation of in-situ doped wrap around contacts with great potential for contact resistance reduction.

Acknowledgment

The imec core CMOS program members, European Commission, local authorities and the imec pilot line are acknowledged for their support. All epitaxial depositions in this work were done on ASM Intrepid® high volume reduced pressure CVD reactor.

References

[1] H. Wu, O. Gluschenkov, G. Tsutsui, C. Niu, K. Brew, C. Durfee, C. Prindle, V. Kamineni, S. Mochizuki, C. Lavoie, E. Nowak, Z. Liu, J. Yang, S. Choi, J. Demarest, L. Yu, A. Carr, W. Wang, J. Strane, S. Tsai, Y. Liang, H. Amanapu, I. Saraf , K. Ryan, F. Lie, W. Kleemeier, K. Choi, N. Cave, T. Yamashita, A. Knorr, D. Gupta, B. Haran, D. Guo, H. Bu, and M. Khare, IEDM Tech. Dig., p.819 (2018)

[2] S-A. Chew, H. Yu , M. Schaekers, S.Demuynck, G.Mannaert, E.Kunnen, E.Rosseel, A.Hikavyy, A. Dangol, K. De Meyer, D. Mocuta and N.Horiguchi*2017 IEEE Int. Interconnect Techn. Conf. (IITC)*, p.1 (2017)

[3] E. Capogreco, H. Arimura, L. Witters, 1A. Vohra, C. Porret, R. Loo, A. De Keersgieter, E. Dupuy, D. Marinov, A. Hikavyy, F. Sebaai, G. Mannaert, L.-A. Ragnarsson, Y. K. Siew, C. Vrancken, A. Opdebeeck, J. Mitard, R. Langer, E. Altamirano Sanchez, F. Holsteyns, S. Demuynck, K. Barla, V. De Heyn, D. Mocuta, N. Collaert, N. Horiguchi., *Symposium on VLSI Technology,* T94 (2019)

[4] Geoffrey Yeap, S.S. Lin, Y.M. Chen, H.L. Shang, P.W. Wang, H.C. Lin, Y.C. Peng, J.Y. Sheu, M. Wang, X. Chen, B.R. Yang, C.P. Lin, F.C. Yang, Y.K. Leung, D.W. Lin, C.P. Chen, K.F. Yu, D.H. Chen, C.Y. Chang, H.K. Chen, P. Hung, C.S. Hou, Y.K. Cheng, J. Chang, L. Yuan, C.K. Lin, C.C. Chen, Y.C. Yeo, M.H. Tsai, H.T. Lin, C.O. Chui, K.B. Huang, W. Chang, H.J. Lin, K.W. Chen, R. Chen, S.H. Sun, Q. Fu, H.T. Yang, H.T. Chiang, C.C. Yeh, T.L. Lee, C.H. Wang, S.L. Shue, C.W. Wu, R. Lu, W.R. Lin, J. Wu, F. Lai, Y.H. Wu, B.Z. Tien, Y.C. Huang, L.C. Lu, Jun He, Y. Ku, J. Lin, M. Cao, T.S. Chang, S.M. Jang, IEDM Tech. Dig., p.879 (2019)

[5] A. Hikavyy, I. Zyulkov, H. Mertens, L. Witters, R. Loo, N. Horiguchi, *Mater. Sci. Semicond. Process.* **70**, 24 (2017)

[6] A. Hikavyy, A. Kruv, T. V. Opstal, B. De Vos, C. Porret, R. Loo, *Semicond. Sci. Technol.* **32**, 114006, (2017)

[7] L. Liebmann, J. Zeng, X. Zhu, L. Yuan, G. Bouche, J. Kye, Symposium on VLSI Technology, p.1 (2016)

[8] J. Zhang, J. Frougier, A. Greene, X. Miao, L. Yu, R. Vega, P. Montanini, C. Durfee, A. Gaul, S. Pancharatnam, C. Adams, H. Wu, H. Zhou, T. Shen, R. Xie, M. Sankarapandian, J. Wang, K. Watanabe, R. Bao, X. Liu, C. Park, H. Shobha, P. Joseph, D. Kong, A. Arceo De La Pena, J. Li, R. Conti, D. Dechene, N. Loubet, R. Chao, T. Yamashita, R. Robison, V. Basker, K. Zhao, D. Guo, B. Haran, R. Divakaruni, H. Bu, IEDM Tech. Dig., p.250 (2019).

[9] S. Barraud, B. Previtali, V. Lapras, C. Vizioz, J.-M. Hartmann, S. Martinie, J. Lacord, M. Cassé, L. Dourthe, V. Loup, G.Romano, N. Rambal, Z. Chalupa, N. Bernier, G. Audoit, A. Jannaud, V. Delaye, V. Balan, O. Rozeau, T. Ernst, M. Vinet, IEDM Tech. Dig., 500 (2018)

[10] H. Mertens, R. Ritzenthaler, Y. Oniki, B. Briggs, B.T. Chan, A. Hikavyy, T. Hopf, G. Mannaert, Z. Tao, F. Sebaai, A. Peter, K. Vandersmissen, E. Dupuy, E. Rosseel, D. Batuk, J. Geypen, G. T. Martinez, D. Abigail, E. Grieten, K. Dehave, J. Mitard, S. Subramanian, L.-Å. Ragnarsson, P. Weckx, D. Jang, B. Chehab, G. Hellings, J. Ryckaert, E. Dentoni Litta, N. Horiguchi, *will be presented at the 2021 Symposium on VLSI Technology and Circuits (VLSI 2021)*

[11] R. Loo, A. Y. Hikavyy, L. Witters, A. Schulze, H. Arimura, D. Cott, J. Mitard, C. Porret, H. Mertens, P. Ryan, J. Wall, K. Matney, M. Wormington, P. Favia, O. Richard, H. Bender, A. Thean, N. Horiguchi, D. Mocuta, and N. Collaert., *ECS J. of Solid State Sci. and Tech.*, **6** (1), P14 (2017)

[12] J-L. Everaert, M. Schaekers, H. Yu, L.-L. Wang, A. Hikavyy, L. Date, J. del Agua Borniquel, K. Hollar, F. A. Khaja, W. Aderhold, A. J. Mayur, J.Y. Lee, H. van Meer, Y.-L. Jiang, K. De Meyer, D. Mocuta, N. Horiguchi, *Symposium on VLSI Technology*, T214 (2017).

[13] J. Margetis, D. Kohen, C. Porret, L. P. B. Lima, R. Khazaka, G. Rengo, R. Loo, J. Tolle, A. Demos, *ECS Transactions* **93** (1), 7, (2019)

[14] T. Tabata, J. Aubin, K. Huet, F. Mazzamuto, *J. Appl. Phys.* **125**, 215702 (2019)

[15] K. H. Chung, and J. C. Sturm. *ECS Transactions*, **6** (1), 401 (2007)

[16] A. Hikavyy, C. Porret, E. Rosseel, A. Milenin, R. Loo, *Semicond. Sci. Technol.* **34**, 074003 (2019)

Selective Epitaxy of Submicron Ge Wire Structures for Photodetectors and Optical Modulators in Si Photonics

Yasuhiko Ishikawa, Kyosuke Noguchi, Mayu Tachibana, Kazuki Kawashita, Ryota Oyamada, Kazuki Motomura, Shuhei Sonoi, Riku Katamawari, and Takeshi Hizawa

Department of Electrical and Electronic Information Engineering, Toyohashi University of Technology, Toyohashi, Aichi 441-8580, Japan

A selective-area epitaxial growth of Ge on the submicron scale is studied using chemical vapor deposition (CVD) on Si in terms of the fabrication of integrated photonic devices operating at the optical communication wavelengths of around 1.55 μm. A mesa structure of Ge on the micron scale, having a vertical pin junction, is effective for fabricating a waveguide-integrated photodetector in the optical receiver. A 3-dB cutoff frequency is obtained up to approximately 50 GHz. For a higher frequency operation, a narrower structure of submicron-wide Ge wire, having a lateral pin junction, is required. Such a wire structure is also important for the application to an electro-absorption optical intensity modulator in the optical transmitter. Structural and optical properties are investigated for a submicron-wide Ge wire on Si selectively grown by CVD.

Introduction

An epitaxial layer of Ge on a Si-on-insulator (SOI) substrate has been applied to a near-infrared photodetector (PD) integrated with a Si optical waveguide (WG) in the Si-based optical receiver [1–10], which operates at the optical communication wavelengths of 1.3–1.6 μm. An electro-absorption optical intensity modulator (EA-MOD) of Ge on SOI has been also studied [11–16], which utilizes the Franz-Keldysh effect, i.e., an enhanced optical absorption under a high electric field at the wavelengths longer than the fundamental absorption edge. The Ge EA-MOD is effective for the optical transmitter to significantly reduce the energy consumption [11] in comparison with Si MODs utilizing the free-carrier plasma dispersion effect. Despite a large lattice mismatch of 4.2% between Ge and Si, a uniform Ge layer is grown epitaxially on a Si (001) surface using a low–high-temperature two-step growth method, where a buffer layer of pure Ge as thin as 50 nm is grown at <400°C, followed by a growth at, typically, 600°C [17]. A post-growth annealing at higher temperatures (800°C–900°C) reduces the threading dislocation density in Ge on the order of 10^7 cm^{-2} or below, realizing high-performance photonic devices. In the device fabrication, a selective-area growth of Ge via chemical vapor deposition (CVD) has been widely used, where a SiO$_2$ layer acts as an ideal mask to prevent the deposition of Ge at undesired areas [17].

In this paper, a WG-PD of a micron-wide Ge mesa structure with a vertical pin junction is firstly described, which shows a 3-dB cutoff frequency up to approximately 50 GHz. Then, a narrower structure of submicron Ge wire, having a lateral pin junction in the underlying Si layer, is discussed for realizing a higher frequency response. Such a wire structure of Ge is also effective for an EA-MOD operating at around 1.55 μm [16] in contrast to the operation at around 1.60 μm in previous Ge EA-MODs [12,13]. In a Ge layer on Si/SOI, a built-in tensile stress is generated by the thermal expansion mismatch between the Ge layer and the thick Si substrate [18,19], while the stress should be relaxed in the narrow Ge wire, inducing a blue shift in the operating wavelength. A widening of the direct bandgap energy of Ge to induce the blue shift is reported, as observed in micro-photoluminescence (PL) spectra.

Micron-Wide Mesa Structure of Ge by Selective Epitaxial Growth on Si: Application to WG-Integrated Vertical pin PDs

Fabrication Procedure

A WG-PD of micron-wide mesa structure of Ge with a vertical pin junction, shown in Figs. 1(a) and 1(b), was fabricated. An advantage in the vertical pin structure lies in the transit length of photo-generated carriers controlled by the growth thickness of Ge layer. The thickness should be, e.g., as small as 1 μm or less in order to achieve the 3-dB cutoff frequency higher than 20 GHz.

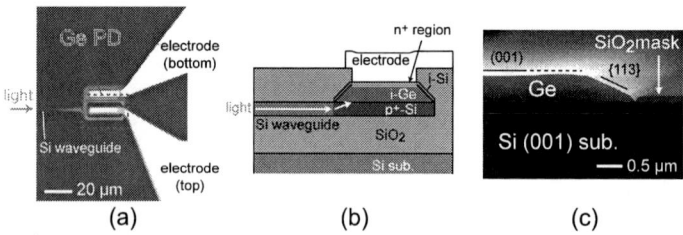

(a)   (b)   (c)

Figure 1. (a) A typical top-view optical microscope image and (b) a schematic cross-section of a vertical pin PD of Ge mesa structure on SOI integrated with a Si optical WG, and (c) a typical cross-sectional scanning electron microscope (SEM) image of Ge mesa structure selectively grown on a reference Si substrate.

As the starting substrate, an SOI wafer with a 220-nm-thick top (001) Si layer was used. First, the top Si layer was patterned to form a channel structure of Si optical WG with the width of 440 nm by electron-beam lithography. At the edge of the WG, a boron-implanted p^+ slab area was prepared, on which the PD was fabricated. The surface was covered with a SiO_2 layer, which acts as the growth mask, while a window area of exposed Si slab surface was partially prepared by photolithography and wet etching. The window size was approximately 8 μm in width and 23 μm in length. The window edges were aligned in the [110] direction. Then, an undoped Ge mesa structure was selectively grown on the Si window surface by ultrahigh-vacuum CVD. 9%-GeH₄/Ar was used as the source gas. A low–high-temperature two-step growth [17] was used to obtain a Ge mesa structure, where a thin buffer layer (~50 nm) of pure Ge was grown at a low temperature of 370°C, followed by a growth of Ge at an elevated temperature of 700°C. The total Ge thickness was designed to be 500 nm. Figure 1(c) shows a typical cross-sectional SEM image of a Ge mesa structure grown on a reference bulk Si wafer. The mesa structure mainly consisted of a top (001) surface and an inclined {113} facet sidewall, which usually appears in the selective growth of Ge [17] because of the growth rate lower than that on the (001) plane. A small mound was also seen on the top surface near the {113} sidewall, probably reflecting the migration of Ge atoms from the {113} sidewall. Then, the Ge surface was covered with a Si cap layer of 50 nm in thickness as a protection layer. A source gas of 10%-Si_2H_6/Ar was used. After a post-growth annealing at 800°C in N_2 to reduce the threading dislocation density in Ge, an n^+ region of the vertical n^+-i-p^+ junction was formed by a phosphorus implantation on the top (001) surface of Ge mesa structure. The phosphorus ions were implanted into the Si cap layer and the upper region of Ge layer near the Si cap layer (approximately 50 nm deep from the Si cap/Ge interface). The in-plane size of the top n^+ region was 5 µm in width and 20 µm in length. Finally, metal electrodes of Al/TiN/Ti were formed.

Current-voltage (I-V) characteristics at 20°C were measured under dark. Temperature dependence (10–80°C) of the I-V characteristics was also measured to discuss the origin of the dark current. Responsivity spectra were measured at 20°C in the wavelength range of the S (1.460–1.530 µm), C (1.530–1.565 µm), and L (1.565–1.625 µm) bands at the reverse bias voltages of 1, 2, and 3 V. A TE-polarized light was introduced through the Si WG. The light was evanescently coupled from the Si WG to the Ge PD because of the refractive index of 4.2 for Ge larger than 3.5 for Si.

PD Characteristics

I-V Curves and Responsivity Spectra. Figure 2(a) shows typical I-V characteristics for a pin PD obtained under dark at different temperatures between 10°C and 80°C. The PD showed good rectifying diode properties. The reverse dark leakage current increased with the temperature, while the current was less than 1 µA even at the increased temperature of 80°C, which is typically required for the practical applications. The Arrhenius plots for the dark current under the reverse biases of 1, 2, and 3 V are shown in Fig. 2(b). The activation energy was around 0.33 eV, the half of the indirect/minimum bandgap energy for Ge. This indicates that the dark current is dominated by the thermal generation of carriers via the mid-gap defect levels in Ge.

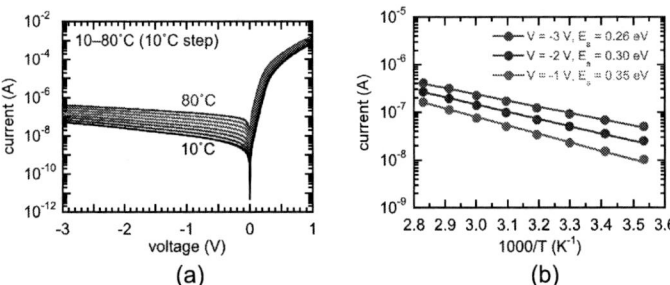

Figure 2. (a) Typical I-V characteristics under dark measured at different temperatures and (b) Arrhenius plots for dark leakage current.

Figure 3(a) shows typical I-V characteristics at 20°C obtained under dark and under the illumination of 1.55-μm light. The dark leakage current at the reverse bias of 1 V showed a small value of 0.02 μA, corresponding to the areal dark current density of 20 mA/cm^2, which is comparable to the lowest value previously reported for a WG-integrated Ge pin PD [1]. Under the illumination, the current increased in the reverse bias, revealing a successful PD operation. The responsivity at 1.55 μm was estimated to be as high as 0.7 A/W, taking into account the light intensity of 33 μW at the PD input, where the propagation loss in the Si WG and the coupling loss between the lensed fiber and the Si WG were removed.

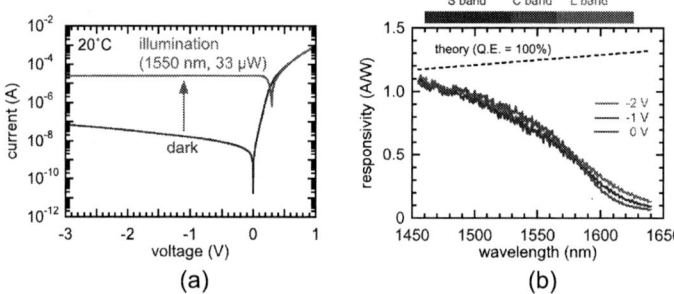

(a) (b)

Figure 3. (a) Typical I-V characteristics with and without the illumination of 1.55-μm light and (b) typical responsivity spectra at different reverse bias voltages of 1, 2, and 3 V (solid lines) with a theoretical one (dashed line) assuming the quantum efficiency (Q.E.) of 100%.

Figure 3(b) shows typical responsivity spectra obtained at different reverse bias voltages of 1, 2, and 3 V. The responsivity at the wavelength of 1.46 μm was as high as 1.1 A/W, which is almost the same as the theoretical one assuming the quantum efficiency of 100%. The responsivity decreased with increasing the wavelength, probably reflecting the reduction in the optical absorption coefficient of Ge, while the responsivity remained as high as 0.6 A/W or larger in the S and C bands. However, the responsivity in the L band should be enhanced for the practical applications, although a small increase in the responsivity with the reverse voltage was observed due to the Franz-Keldysh effect [20].

High-Frequency Response. The blue line in Fig. 4(a) shows a typical frequency response for the PD described above. The PD revealed a 3-dB cutoff frequency as high as 22 GHz. The theoretical 3-dB cutoff frequency is shown in Fig. 4(b) as a function of the i-Ge thickness. In the present PD, the i-Ge thickness was around 0.45 μm. In this case, the theoretical 3-dB cutoff frequency is as high as 50 GHz, taking into account the RC delay in addition to the carrier transit delay, while assuming no parasitic series resistance in the PD. The obtained frequency of 22 GHz is significantly smaller than the theoretical one. This is probably ascribed to an increase in the RC delay induced by a series resistance. In fact, the series resistance was estimated to be as high as 200 Ω based on the static I-V curve under the forward bias shown in Fig. 3(a), and the theoretical one in Fig. 4(b), taking into account the series resistance of 200 Ω, showed a good agreement with the observed value of 22 GHz.

A reduction in the series resistance led to the 3-dB cutoff frequency as high as 48 GHz, as shown by the red line in Fig. 4(a). However, it is not easy to further increase the frequency in the vertical pin PD. This is because the relatively large capacitance/junction area prevents the reduction in the RC delay even if the series resistance is reduced. A narrower structure of submicron Ge wire, having a lateral pin junction in the underlying Si layer, is favorable for a higher frequency operation.

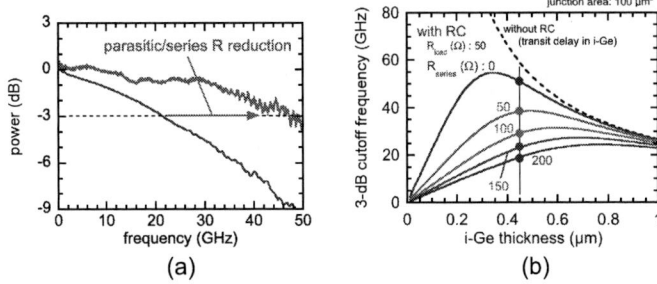

Figure 4. (a) Typical frequency responses and (b) theoretical 3-dB cutoff frequencies as a function of the i-Ge thickness.

Submicron Wire Structure of Ge toward Higher-Performance PDs and EA-MODs

Advantages of Submicron Wire Structure of Ge

For enhancing the operating frequency of >50 GHz with keeping a high responsivity, a submicron-wide Ge wire structure with a lateral pin junction is effective because of the RC delay smaller than that for the conventional vertical pin PD [21,22]. In comparison with a vertical pin PD schematically shown in Fig. 5(a), a lateral pin PD in Fig 5(b) can reduce the junction capacitance due to a smaller junction area, meaning that the RC delay is efficiently reduced. Here, the transit delay dominates the operating frequency, which depends on the distance between the p and n regions. In order to achieve the operating frequency of >50 GHz, the intrinsic region between the p and n regions should be as small as 0.5 μm or less, according to Fig. 4(b). The preparation of Ge wire structure with the submicron width is obviously important.

In turn, in a submicron wire structure, the built-in tensile stress, generated by a thermal expansion mismatch between the Ge layer and the Si substrate [18,19], should be reduced by an elastic relaxation of lattice strain due to a structural deformation. The narrowed direct bandgap of 0.77 eV under the tensile stress should be widened under the reduced stress, approaching to the unstrained case of 0.80 eV. This should induce a blue shift in the fundamental optical absorption edge from 1.61 μm to ~1.55 μm in wavelength, leading to a reduction in the responsivity for PD in the C band in addition to the L band. A longer PD would be required to maintain the high responsivity, while preventing a significant increase in the capacitance, or the RC delay. On the other hand, for Ge EA-MOD based on the Franz-Keldysh effect, this blue shift is favorable for realizing the operating wavelength at around 1.55 μm, or the C band. In previous studies, the EA-MODs of Ge only worked in the longer wavelength range of the L band [12,13], i.e., for the C band operation, a SiGe alloy was required for increasing the bandgap [11]. It is noted that the structural difference between the PD and EA-MOD of a submicron Ge wire is only the presence of an output Si WG in the EA-MOD except for the Ge length, while the cross-sectional structure of the Ge wire can be easily designed to act as a single-mode WG required for the EA-MOD application.

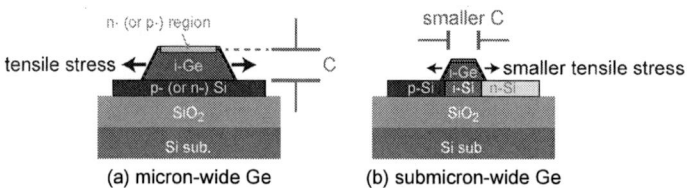

(a) micron-wide Ge   (b) submicron-wide Ge

Figure 5. Schematic illustrations of (a) micron-wide Ge PD with a vertical pin junction and (b) submicron-wide Ge PD with a lateral pin junction in the underlying Si layer.

Selective Epitaxy of Submicron Wire Structure of Ge on Si

In terms of higher-performance PDs and EA-MODs, a selective epitaxial growth of submicron-wide Ge wire was carried out on a (001) Si wafer, on which a SiO_2 layer was thermally grown with a thickness of 100 or 200 nm. First, the SiO_2 layer was partially removed to open the strip-shaped windows of the exposed Si surface by i-line photolithography and wet etching. The width of the strip window was changed as a parameter (the minimum width of 0.7±0.1 μm), while the length was sufficiently long (>1 mm). The window edges were aligned in the [110] direction. Similar to the case of the micron-wide Ge mesa in the previous section, a Ge epitaxial layer was selectively grown on the exposed Si windows via UHV-CVD with the two-step growth method (370°C/700°C). The thickness of Ge was designed to be 500 nm on the macroscopic (001) Si surface. Then, a Si cap layer of 50 nm in thickness (on the (001) plane) was grown on Ge for passivating the chemically unstable surface of Ge.

Figure 6(a) shows a typical cross-sectional SEM image of a Ge wire structure without a Si cap layer. The surface was mainly composed of {113} and {111} facet planes, while a top (001) plane with a width as narrow as 0.1 μm remained. The height of approximately 370 nm is smaller than 500 nm exhibited on the macroscopic (001) plane. Figures 6(b) shows a typical cross-sectional SEM images of the Ge wire with a Si cap layer grown at 600°C. A roughened surface was observed on the inclined facet planes in contrast to the case of the micron-wide mesa structure. Depressions were specifically induced on the {113} planes near the boundary with the {111} planes. The depressions disappeared when the capping temperature was reduced to 530°C, as shown in Fig. 6(c). An almost uniform capping of Si was realized, being conducive to the fabrication of practical devices, although the thickness was as low as 30 nm on the wire surfaces of the {113} and {111} planes (c.f., 50 nm on the (001) surface). The reduced Si thickness on the facet planes is attributed to the lower growth rates of Si than that on the (001) plane. The details on the formation of the Si cap layer are described elsewhere [23].

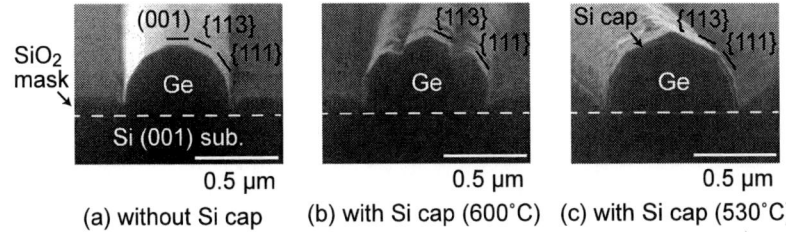

(a) without Si cap   (b) with Si cap (600˚C)   (c) with Si cap (530˚C)

Figure 6. Typical cross-sectional SEM images of Ge wire structures with and without a Si cap layer. (a) An uncapped wire, (b) a Si-capped wire in the case of the capping temperature of 600°C, and (c) a Si-capped wire in the case of the capping temperature of 530°C. The sample in (c) was dipped into a 3% H_2O_2 solution before the observation to enhance the image contrast, resulting from a selective etching of Ge by several 10 nm.

Bandgap Engineering

Typical micro-PL spectra are shown in Fig. 7, which was measured using an excitation laser source with a wavelength of 785 nm (the nominal $1/e^2$ spot diameter of 2 μm). For the 0.9-μm-wide Ge wire structure, the PL peak due to the direct transitions was located at the wavelength of about 1.54 μm, which is shorter than 1.57 μm for the 10-μm-wide Ge mesa on Si. The peak position for the wire is similar to that for a bulk Ge wafer, i.e., there is a blue shift for the wire, corresponding to a widening of direct bandgap. The blue shift/bandgap widening probably results from a relaxation of grown-in tensile strain. As described earlier, the blue shift plays a significant role for Ge EA-MODs. In contrast to the operating wavelength range of >1.60 μm in previous studies [12,13], an EA-MOD of a submicron Ge wire favorably operates in the 1.55 μm range of the C band, as reported recently [16].

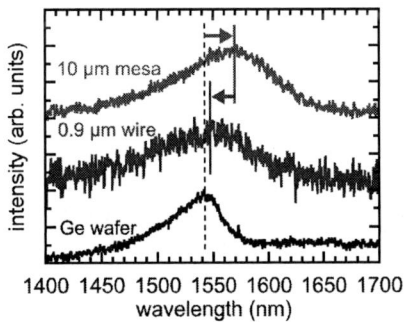

Figure 7. Typical PL spectra for a 10-μm-wide mesa (top), a 0.9-μm-wide wire (middle), and a bulk Ge wafer (bottom).

Summary

Ge epitaxial layers were selectively grown by CVD on Si from the viewpoint of photonic devices in Si photonics operating at around 1.55 μm. A micron-wide mesa structure of Ge with a vertical pin junction is effective for a WG-integrated PD with a 3-dB cutoff frequency up to approximately 50 GHz. A narrower structure of submicron Ge wire, having a lateral pin junction in the underlying Si layer, is required for a higher frequency operation. Such a wire structure is also effective for an EA-MOD operating in the C band.

Acknowledgments

The authors would like to thank K. Inaba and M. Shimokawa for their supports in experiments. The authors would also like to thank Prof. K. Wada of The University of Tokyo/Massachusetts Institute of Technology, Dr. H. Fukuda, Dr. T. Hiraki, Dr. T. Tsuchizawa, and Dr. Y. Maeda of NTT Device Technology Laboratories for valuable discussions on Ge PDs, and Dr. J. Fujikata of PETRA for valuable discussions on Ge wire structures. This work was partly supported by "R&D on optical PLL device for receiving and monitoring optical signals", the Commissioned Research of National Institute of Information and Communication Technology, Japan (Grant Number 18101), by JSPS KAKENHI Grant Number JP21H01367, and by Casio Science Promotion Foundation.

References

1. D. Ahn, C. -Y. Hong, J. Liu, W. Giziewicz, M. Beals, L. C. Kimerling, and J. Michel, Opt. Express **15**, 3916 (2007).
2. T. Yin, R. Cohen, M. M. Morse, G. Sarid, Y. Chetrit, D. Rubin, and M. J. Paniccia, Opt. Express **15**, 13965 (2007).
3. G. Masini, S. Sahni, G. Capellini, J. Witzens, and C. Gunn, Adv. Opt. Technol. **2008**, 196572 (2008).
4. L. Vivien, J. Osmond, J. -M. Fédéli, D. Marris-Morini, P. Crozat, J. -F. Damlencourt, E. Cassan, Y. Lecunff, and S. Laval, Opt. Express **17**, 6252 (2009).
5. K. W. Ang, T. Y. Liow, M. B. Yu, Q. Fang, J. Song, G. Q. Lo, and D. L. Kwong, IEEE J. Sel. Top. Quantum Electron. **16**, 106 (2010).
6. N. -N. Feng, P. Dong, D. Zheng, S. Liao, H. Liang, R. Shafiiha, D. Feng, G. Li, J. E. Cunningham, A. V. Krishnamoorthy, and M. Asghari, Opt. Express **18**, 96 (2010).
7. S. Park, T. Tsuchizawa, T. Watanabe, H. Shinojima, H. Nishi, K. Yamada, Y. Ishikawa, K. Wada, and S. Itabashi, Opt. Express **18**, 8412-8421 (2010).
8. H. Nishi, T. Tsuchizawa, R. Kou, H. Shinojima, T. Yamada, H. Kimura, Y. Ishikawa, K. Wada, and K. Yamada, Opt. Express **20**, 9312 (2012).
9. T. Hiraki, H. Nishi, T. Tsuchizawa, R. Kou, H. Fukuda, K. Takeda, Y. Ishikawa, K. Wada, and K. Yamada, IEEE Photonics J. **5**, 4500407 (2013).
10. K. Ito, T. Hiraki, T. Tsuchizawa, and Y. Ishikawa, Jpn. J. Appl. Phys. **56**, 04CH05 (2017).

11. J. Liu, M. Beals, A. Pomerene, S. Bernardis, R. Sun, J. Cheng, L. C. Kimerling, and J. Michel, Nature Photon. **2**, 433 (2008).
12. A. E. -J. Lim, T. -Y. Liow, F. Qing, N. Duan, L. Ding, M. Yu, G. -Q. Lo, and D. -L. Kwong, Opt. Express **19**, 5040 (2011).
13. N. -N. Feng, D. Feng, S. Liao, X. Wang, P. Dong, H. Liang, C. -C. Kung, W. Qian, J. Fong, R. Shafiiha, Y. Luo, J. Cunningham, A. V. Krishnamoorthy, and M. Asghari, Opt. Express **19**, 7062 (2011).
14. S. A. Srinivasan, M. Pantouvaki, S. Gupta, H. T. Chen, P. Verheyen, G. Lepage, G. Roelkens, K. Saraswat, D. Van Thourhout, P. Absil, and J. Van Campenhout, J. Lightwave Technol. **34**, 419 (2016).
15. L. Mastronardi, M. Banakar, A. Z. Khokhar, N. Hattasan, T. Rutirawut, T. Dominguez Bucio, K. M. Grabska, C. Littlejohns, A. Bazin, G. Mashanovich, and F. Y. Gardes, Opt. Express **26**, 6663 (2018).
16. J. Fujikata, M. Noguchi, K. Kawashita, R. Katamawari, S. Takahashi, M. Nishimura, H. Ono, D. Shimura, H. Takahashi, H. Yaegashi, T. Nakamura, and Y. Ishikawa, Opt. Express **28**, 33123 (2020).
17. H. -C. Luan, D. R. Lim, K. K. Lee, K. M. Chen, J. G. Sandland, K. Wada, and L. C. Kimerling, Appl. Phys. Lett. **75**, 2909 (1999).
18. Y. Ishikawa, K. Wada, D. D. Cannon, J. Liu, H. -C. Luan, and L. C. Kimerling, Appl. Phys. Lett. **82**, 2044 (2003).
19. Y. Ishikawa, K. Wada, D. D. Cannon, J. Liu, H. -C. Luan, J. Michel, and L. C. Kimerling, J. Appl. Phys. **98**, 013501 (2005).
20. K. Takeda, T. Hiraki, T. Tsuchizawa, H. Nishi, R. Kou, H. Fukuda, T. Yamamoto, Y. Ishikawa, K. Wada, and K. Yamada, IEEE J. Sel. Top. Quantum Electron. **20**, 3800507 (2014).
21. S. Sahni, N. K. Hon, and G. Masini, ECS Trans. **64**, 783 (2014).
22. N. K. Hon, S. Sahni, A. Mekis, and G. Masini, *14th International Conference on Group IV Photonics (GFP2017)* (IEEE, 2017), p. 177.
23. R. Katamawari, K. Kawashita, T. Hizawa, and Y. Ishikawa, to be published in J. Vac. Sci. Technol. **39**, (2021)

Fabrication of Ge-on-Insulator by epitaxial growth and ion-implanted exfoliation for electronics and optoelectronics applications

Keisuke Yamamoto[a], Dong Wang[a], and Hiroshi Nakashima[b]

[a] Faculty of Engineering Sciences, Kyushu University, Kasuga, Fukuoka 816-8580, Japan
[b] Global Innovation Center, Kyushu University, Kasuga, Fukuoka 816-8580, Japan

Ge has been received much interest as a CMOS channel material and a near-infrared optical material due to its superior characteristics. Ge-on-Insulator (GOI) structure is necessary to suppress large leakage current originating from the narrow bandgap for application use. A method combined wafer bonding and layer splitting by hydrogen ion (H^+) implantation, known as Smart-Cut[TM] realized for Si-on-Insulator fabrication, has been tried to apply for fabricating GOI with large diameter, uniform thickness, and single crystal. In this study, we fabricated GOI by Smart-Cut[TM] technique and demonstrated electronic and optoelectronic devices on the GOI. Besides, we combined a Ge epitaxial growth method with the Smart-Cut[TM] technique to improve GOI quality.

1. Introduction

Many researchers study novel material introduction instead of Si energetically for continuous performance improvement of CMOS and related devices. Ge has received much interest as a CMOS channel material due to its superior characteristics, particularly high carrier mobilities (1). Besides, the quasi-direct band structure of Ge enables a certain population in Γ valley due to intervalley scattering of conduction band electrons, and it can be applied for near-infrared optical applications such as on-chip optical interconnect (2,3). On the other hand, Ge has a drawback of relatively large leakage current density due to its narrow bandgap. To enjoy the superior characteristics of Ge and suppress the leakage current density, Ge-on-Insulator (GOI) structure has been proposed, which consists of a single crystalline Ge thin layer formed on a buried oxide (BOX). Although GOI fabrication is challenging, many research groups have tried to form high-quality GOI and suggested several methods. For example, SiGe condensation, (4,5) direct wafer bonding, (6,7) epitaxial growth on a mother substrate and layer transfer, (8,9) and layer splitting by hydrogen ion (H^+) implantation, which is called Smart-Cut[TM]. Smart-Cut[TM] is maturely used in Si-on-Insulator fabrication and has been tried to apply for GOI with large diameter and uniform thickness (10-13). However, electrical characterizations of GOI and demonstrations of electronics and near-infrared optical devices on GOI are limited (2,14).

In this paper, we first show the electrical characteristics of Smart-Cut[TM] GOI fabricated from a bulk p-Ge wafer. We also demonstrate operations of p- and n-channel MOSFETs with metal source/drain (S/D) fabricated on the GOI. Second, we demonstrate the operation of an asymmetric metal/Ge/metal light-emitting diode on the GOI. Compared to the same device fabricated on bulk Ge, the electroluminescence (EL) intensity of the GOI diode was drastically improved owing to the carrier confinement effect originated from the small thickness of GOI. Third, the difficulty in fabricating n-type GOI is demonstrated. From the structural analysis, the difficulty is originated from the remaining defect introduced during H^+ ion implantation. We also propose a novel method to avoid these defects and succeeded in the fabrication of GOI substrated with n-type conduction and good electron mobilities.

2. GOI fabrication by conventional Smart-Cut™

3.1 Sample preparation procedure and the results of Hall effect measurement

Figure 1 shows the fabrication procedure for GOI in this study (10, 13). In Fig. 1, specs of a donor Ge and a handle Si substrates are also listed. H^+ ion implantation, with a dose of 4×10^{16} cm^{-2} and acceleration energy of 120 keV, into Ge was performed through 100 nm-thick SiO$_2$. After SiO$_2$ removal, 3 nm-thick Al$_2$O$_3$ was deposited on the Ge using atomic layer deposition (ALD). For the Si handle substrate, 50 nm-thick SiO$_2$ was grown by thermal oxidation at 1000 °C. Then, the Al$_2$O$_3$/Ge and the SiO$_2$/Si were manually bonded in the cleanroom atmosphere. To enhance bonding strength, annealing was performed in N$_2$ ambient at 300 °C (15). Continuously, annealing temperature increased to 400 °C, and layer splitting was carried out. The GOI was thinned by wet etching using dilute H$_2$O$_2$ solution, and its surface was flattened by chemical mechanical polishing (CMP). For electrical properties improvement, the GOI was annealed at 500 °C for 1 h in N$_2$ ambient.

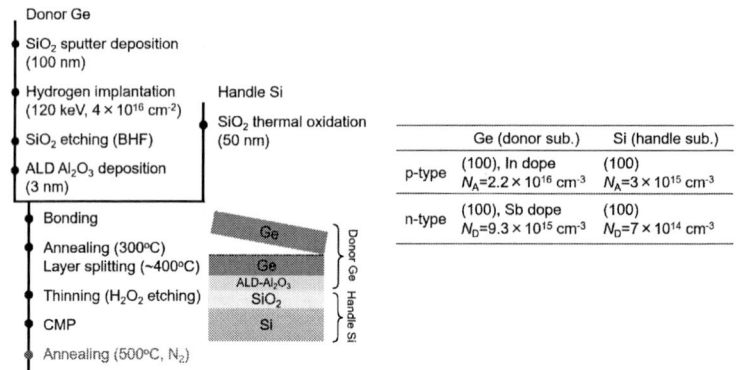

Figure 1. Fabrication procedure of GOI in this study (conventional Smart-Cut™).

Firstly, we discuss the GOI results formed from a bulk p-type Ge as a donor material. Figures 2(a) and 2(b) show hole concentration and hole mobility measured by the Hall effect for the p-GOIs, respectively. Hole concentration decreased, and mobility increased after annealing at 500 °C. This result well agrees with Ref. 10. Figure 3 shows the dependence of hole mobility and concentration on the GOI thickness. Initially, we prepared the GOI with a thickness of 540 nm and carried out a Hall effect measurement. Then, the GOI layer was thinned by dilute H$_2$O$_2$ etching and measured again. This cycle was performed several times for the same sample. We can find that its mobility decreases with decreasing GOI thickness. Also, the hole concentration increases with decreasing GOI thickness. Both values change drastically when the GOI thickness is thinner than 150 nm.

Improvement of hole mobility and concentration can be explained by defect recovery after the GOI formation. In the Ge layer after layer splitting, it is naturally conjectured that there are many defects caused by H^+ ion implantation. Thus the electrical property of as-formed GOI is insufficient compared with the original bulk Ge. These defects may recover by annealing and the electrical property improved when GOI thickness is relatively thick (> 150 nm).

The GOI results formed from a bulk n-type Ge as a donor substrate are described in section 3.

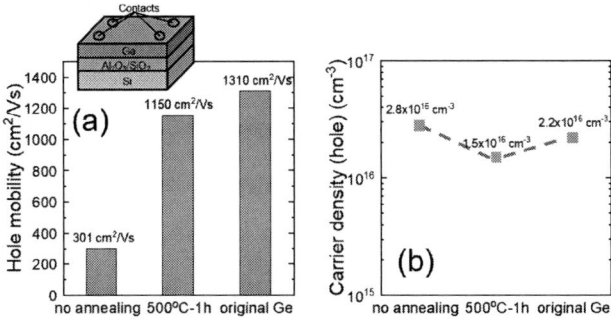

Figure 2. (a) Hole mobility and (b) concentration of the fabricated GOI before and after annealing. GOI thickness is 450 nm. For comparison, the values of original bulk p-Ge (donor material) are also shown.

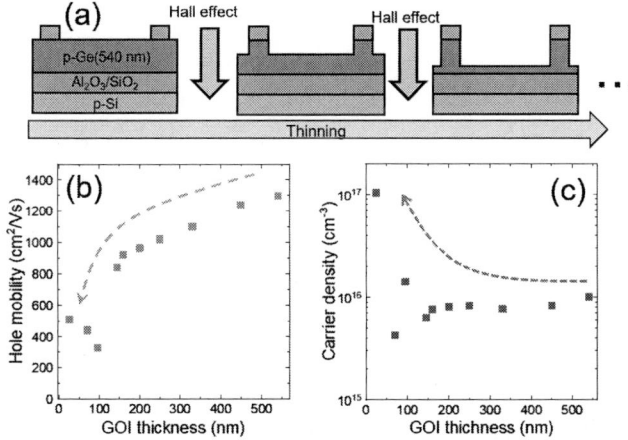

Figure 3. (a) Experimental scheme of cyclic GOI thinning and Hall effect measurement. GOI thickness dependence for (b) hole mobility and (c) hole concentration.

3.2 Demonstration of p- and n-MOSFET with metal S/D on the GOI

We fabricated p- and n-MOSFET with metal S/D on the GOI. Figure 4(a) shows the device fabrication procedure. Here, p-MOSFET is an accumulation channel device with an ohmic PtGe S/D, and n-MOSFET is an inversion channel device with a rectifying TiN S/D (16,17). After GOI formation including final annealing, an active region was defined by island shape etching of the GOI. Here, GOI thicknesses are 100 nm and 150 nm for p-MOSFET and n-MOSFET, respectively. Then metal S/D electrodes were formed. For p-MOSFET, a Pt was deposited by sputtering, and subsequent annealing changed it to PtGe. For n-MOSFET, a TiN that has a low electron barrier height for p-Ge was deposited by sputtering. After sample cleaning, a SiO_2/GeO_2 stacked gate insulator with a thickness of ~ 50 nm was formed by PVD. A thermal evaporated Al was used for the gate and the contact electrodes. Figures 4(b) and 4(c) show the I_D-V_D characteristics for the p-MOSFET and the n-MOSFET, respectively. Both devices successfully operated as an accumulation and an inversion mode FETs. The calculated peak field-effect mobilities for p- and n-MOSFETs are 150 cm^2/Vs and 100 cm^2/Vs, respectively. Although high leakage current must be suppressed, these results can open a way for metal S/D Ge CMOS on GOI substrate.

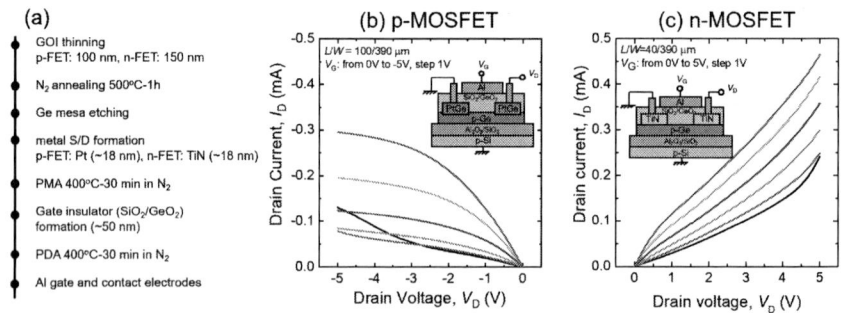

Figure 4. (a) Fabrication flow of metal S/D Ge MOSFET on GOI. I_D-V_D characteristics of the devices. (b) Accumulation mode p-MOSFET with PtGe S/D. (c) Inversion mode n-MOSFET with TiN S/D.

3. Electroluminescence of asymmetric metal/Ge/metal diode on GOI

We fabricated an asymmetric metal/Ge/metal diode to investigate EL performance. The fabrication process is briefly introduced in the following, of which the details are given elsewhere (3). We used a GOI substrate with a four-layer structure of 500 nm-Ge/3 nm-Al_2O_3 /100 nm-SiO_2/Si handle substrate. Here, p-type bulk Ge was used for the GOI fabrication. Firstly, Ge islands (length = 30 μm, width = 30 μm) as active regions were formed by chemical etching (NH_3 : H_2O_2 : H_2O = 1 : 7 : 40 mixed solution). Then, 20 nm-Ti/30 nm-Pt layers and TiN layer were deposited asymmetrically as hole and electron injection electrodes, respectively. All metal layers were deposited using an RF magnetron sputtering method and patterned using a lift-off process. After post-metallization annealing in N_2 at 400 °C for 30 min, the sample surface was passivated using SiO_2/GeO_2

stacked layers, followed by post-deposition annealing in N_2 at 385 °C for 30 min. Finally, Al electrodes were formed using thermal evaporation followed by a contact annealing at 300 °C for 10 min. For comparison, we also fabricated the same diodes on a bulk p-Ge substrate with a slightly modified fabrication process. The active region area is 300 μm^2 for both devices fabricated on GOI and bulk Ge substrates.

Figures 5(a) and 5(b) show EL spectra for the fabricated devices diodes on bulk p-Ge and p-GOI, respectively. The EL measurement was carried out at room temperature under various injection current intensities from 10 to 50 mA. The EL peaks at around 0.8 eV are clearly observed in both diodes, corresponding to the direct bandgap (DBG) energy of Ge. The EL intensity (I_{EL}) of the p-GOI device is higher than that of the bulk p-Ge one (note the different vertical axis scales). The EL spectra of the p-GOI diode also show broad peaks.

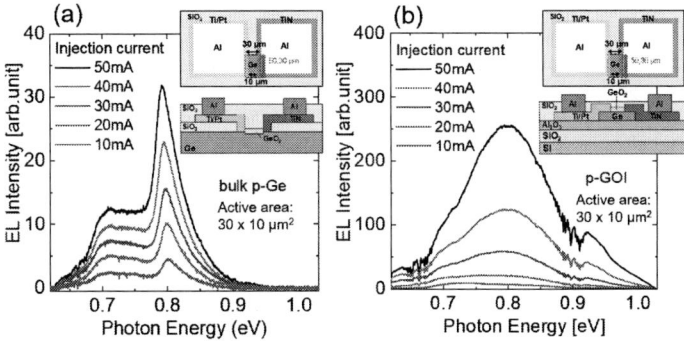

Figure 5. EL spectra for (a) bulk p-Ge and (b) p-GOI diodes with an active region of 30 $\mu m \times 10 \mu m$ (width × length). Copyright (2019) The Japan Society of Applied Physics (3).

One of the possible reasons for the I_{EL} enhancement of GOI is the reflection at the Ge/BOX interface (18). However, even if all the reflected EL beams at Ge/BOX interface can transmit through the top-SiO2/Ge interface (practically impossible), the reflection only gives a maximum enhancement of 50% in EL intensity, which is far from the experimental result as shown in Fig. 5. Therefore, there is another reason for I_{EL} enhancement on GOI.

We believe the different characteristics of the two devices in Fig. 5 are mainly caused by the carrier confinement in the thin GOI layer. First, the integrated EL intensity is always lower for bulk p-Ge devices compared with p-GOI devices. This can be explained by the difference in the running depth of injected holes and electrons in the bulk Ge. Injected holes and electrons diffuse to the depth direction during their transit time. This diffusion causes decreased carrier density in the active region. In addition, injected holes will be repulsed by the positive fixed charges in the passivation insulator and tend to go to deeper positions in the active region. On the other hand, the electrons will be attracted by the positive fixed charges and tend to go near to the passivation layer/Ge interface (the shallow positions of GOI). This charges distribution difference means fewer electrons and holes can meet with each other to accomplish radiative recombination. The problems mentioned above become less severe in GOI diodes. The thin Ge layer of GOI confines carriers in the depth direction and maintains a high contribution of injected carriers to radiative recombination. Therefore, the p-GOI diodes always show higher EL intensity than the bulk p-Ge diodes.

Different from bulk p-Ge diodes, the EL spectra for p-GOI diodes show broad profiles. To investigate this, we performed micro photoluminescence (μ-PL) measurement at room temperature by using a continuous wave laser with a wavelength of 532 nm, a power of 10 mW, and a spot diameter of fewer than 2 μm on the sample surface, as shown in Fig. 6(a). Figures 6(b) and 6(c) show both the μ-PL and EL spectra for bulk p-Ge and p-GOI diodes, respectively. The EL spectra were measured under a current intensity of 30 mA. As for the bulk p-Ge device, the dependence of the EL spectrum on measurement position is small. Therefore, only one EL spectrum of bulk p-Ge was plotted in Fig. 6(b). In addition, the EL and PL spectra show very similar profiles for the bulk p-Ge sample. On the other hand, the μ-PL spectra of the p-GOI sample varied depending on the measurement position. The DBG (0.8 eV) peak intensity decreased with the increasing intensity of peaks at an energy below the bandgap. These low-energy peaks also showed varied peak positions, implying their defect nature. The μ-PL results implied apparent fluctuation of defect type and density in p-GOI. Most of the defects must be generated during the p-GOI fabrication process, and we believe these defects originated from H$^+$ ion implantation. As for the EL spectra, defect-related EL signals with varied peak positions were collected from all the active regions (as shown in Fig. 6(a)), resulting in a broad profile at the low-energy part.

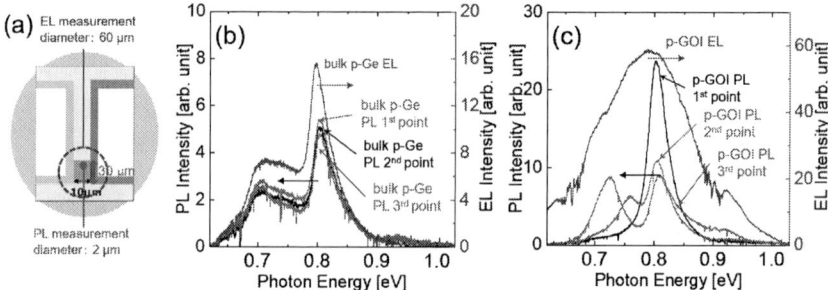

Figure 6. (a) Schematic image of EL measurement area and PL measurement spot in this study. PL and EL spectra for (b) bulk p-Ge and (c) p-GOI diodes with an active region of of 30 μm × 10 μm (width × length). Copyright (2019) The Japan Society of Applied Physics (3).

4. Fabrication and quality improvement of n-type GOI

4.1 Electrical characteristics for GOIs fabricated from bulk n-Ge as donor substrate

In the previous sections, we showed GOI fabrication by conventional Smart-Cut™ and demonstrated the operation of MOSFETs and light emitting diodes fabricated on GOI substrates. Although the GOI fabrication and the device's operations were succeeded, there are remaining problems of residual defect that may be introduced during H^+ ion implantation to the donor Ge and cannot be completely recovered by annealing. In addition, for DBG EL, n-type Ge is preferable because higher electron concentration in Γ valley owing to intervalley scattering enhances the EL intensity (19). Therefore, the formation of n-type GOI with low defect density is desired. In this section, first, we show the trial of n-GOI fabrication by the conventional Smart-Cut^{TM} method. Unfortunately, it is impossible, and we observed residual defects in GOI from TEM analysis. Second, we propose a novel method to avoid the defects and succeeded in making n-GOI.

We fabricated n-GOI with the same method described in section 2 (Fig. 1). Figures 7(a) and 7(b) show carrier concentration and mobility measured by the Hall effect, respectively. Although the GOI layer showed n-type conduction before annealing at 500 °C, annealing at 500 °C changed it to p-type. Similar conduction type change has been reported for H^+ implanted bulk Ge (20). To investigate the reason, a cross-sectional TEM analysis was carried out. Fig. 7(c) shows the TEM image after annealing. We can observe many defects that appear as black dots near the Ge/BOX interface, even after annealing. The defects near the BOX are serious because this region will be kept after GOI thinning. It is well known that defects in Ge, especially vacancy, act as acceptors (21). Therefore, elimination of these defects is essential to maintain n-type conduction, as well as to improve the quality of p-GOI. Since this defective region was just below the top surface of Ge during H^+ implantation, we believe these defects were introduced by H^+ implantation.

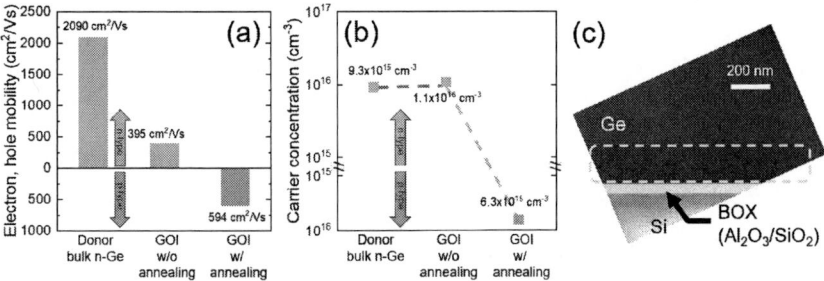

Figure 7. (a) Carrier mobility and (b) concentration of the fabricated GOI before and after annealing. GOI thickness is 270 nm. For comparison, the values of original bulk n-Ge (donor material) are also shown. (c) Cross-sectional TEM image of the GOI after 500 °C-1h annealing. Copyright (2019) The Electrochemical Society (13).

4.2 Novel method to reduce defects for GOI

To reduce defect introduction during GOI fabrication, we propose two novel methods. The first one is to remove the above-mentioned damaged Ge layer by wet etching using dilute H_2O_2 before bonding, as shown in Fig. 8(a). In this method, because the severely damaged Ge layer in the donor Ge wafer was removed in advance, only a less damaged layer was transferred to the handle Si substrate. Another method is the epitaxial growth of a new n-Ge layer on the donor Ge substrate after H^+ implantation, as shown in Fig. 9(a). In this method, all defective layer was removed during GOI thinning after layer splitting, and only epi-Ge layer remains on the handle Si. Figures 8(b) and 9(b) show electron concentrations and mobilities of the GOIs fabricated by these alternative methods. Both methods successfully maintain the conduction type of GOI after annealing at 500 °C. These results suggest that defects elimination before bonding is essential for GOI fabrication using Smart-Cut[TM] technology. We believe these methods also improve the quality of Smart-Cut[TM] p-type GOI substrates.

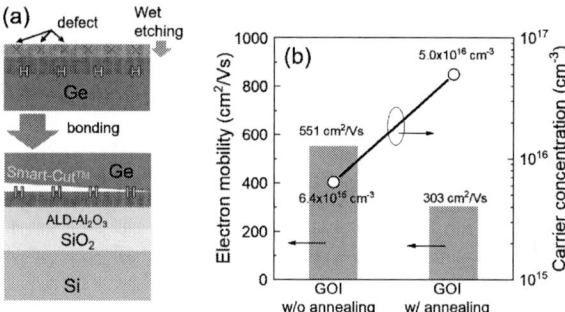

Figure 8. (a) n-GOI fabrication by defect removal (wet etching) and (b) its electrical characteristics. GOI thickness is 230 nm. Copyright (2019) The Electrochemical Society (13)

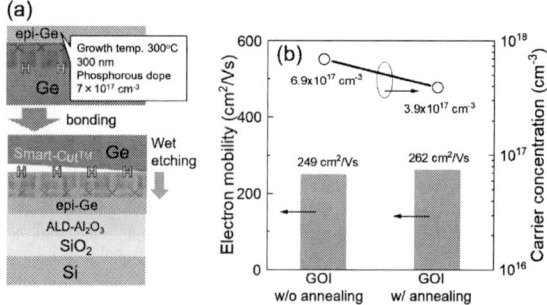

Figure 9. (a) Avoiding defects by epitaxial growth and (b) its electrical characteristics. GOI thickness is 195 nm. Copyright (2019) The Electrochemical Society (13).

5. Conclusion

In this study, we characterized GOI substrates fabricated by the Smart-Cut[TM] technique. Annealing effectively improved the electrical characteristics of a GOI substrate made from bulk p-Ge as donor material. The metal S/D p- and n-MOSFETs fabricated on the GOI substrate successfully operated in both accumulation and inversion modes, respectively, though there is a lot of room for improvement in reducing leakage current density under off-state. An asymmetric metal/Ge/metal light-emitting diode fabricated on the GOI achieved higher EL intensity than the same device fabricated on bulk Ge, which can be explained by the carrier confinement effect due to the small thickness of the GOI layer. Meanwhile, the GOI diode showed a broader EL spectrum than that of bulk Ge. These disadvantages (leakage current of the MOSFETs and broad EL spectrum) mean that the GOI is relatively defective. To reduce these defects and achieve n-type conduction, we propose two alternative ways to fabricate GOI, i.e., to eliminate the defective layer caused by H^+ implantation before bonding. Preliminary results show that n-type conduction was successfully maintained with good electron mobilities even after annealing at 500 °C.

Acknowledgments

This work was partially supported by (JSPS) KAKENHI (grant numbers 19K15028 and 17H03237), MEXT/JSPS Leading Initiative for Excellent Young Researchers (LEADER), QR program, Kyushu University, and Cooperative Research Project Program of the Research Institute of Electrical Communication, Tohoku University.

References

1. A. Toriumi and T. Nishimura, *Jpn. J. Appl. Phys.*, **57**, 010101 (2018).
2. J. Kang, S. Takagi, and M. Takenaka, *Opt. Express*, **26**, 30546 (2018).
3. T. Maekrua, T. Goto, K. Nakae, K. Yamamoto, H. Nakashima, and D. Wang, *Jpn. J. Appl. Phys.*, **58**, No. SB, SBBE05 (2019), 10.7567/1347-4065/aafb5e.
4. S. Nakaharai, T. Tezuka, N. Sugiyama, Y. Moriyama, and S. Takagi, *Appl. Phys. Lett.* **83** 3516 (2003).
5. W. K. Kim, K. Kuroda, M. Takenaka, and S. Takagi, *IEEE Trans. Electron Devices*, **61** 3379 (2014).
6. Z. Zheng, X. Yu, M. Xie, R. Cheng, R. Zhang, and Y. Zhao, *Appl. Phys. Lett.*, **109** 023503 (2016).
7. M. Veerappan, A. Mukannan, F. Salleh, Y. Shimura, Y. Hayakawa, and H. Ikeda, *Semicond. Sci. Technol.*, **32** 035021 (2017).
8. T. Maeda, W.-H. Chang, T. Irisawa, H. Ishii, H. Hattori, V. Poborchii, Y. Kurashima, H. Takagi, and N. Uchida, *Appl. Phys. Lett.*, **109** 262104 (2016).
9. K. Sawano, Y. Hoshi, S. Endo, T. Nagashima, K. Arimoto, J. Yamanaka, K. Nakagawa, S. Yamada, K. Hamaya, M. Miyao, and Y. Shiraki, *Thin Solid Films*, **557** 76 (2014).
10. J. Kang, X. Yu, M. Takenaka, and S. Takagi, *Mater. Sci. Semicond. Process.*, **42** 259 (2016).
11. M. Kim, S. J. Cho, Y. J. Dave, H. Mi, S. Mikael, J.-H. Seo, J. U Yoon, and Z. Ma, *Semicond. Sci. Technol.*, **33**, 015017 (2018).
12. Y. Ruan, R. Liu, W. Lin, S. Chen, C. Li, H. Lai, W. Huang, and X. Zhang, *J. Electrochem. Soc.*, **158**, H1125 (2011).
13. K. Yamamoto, K. Nakae, H. Akamine, D. Wang, H. Nakashima, M. M. Aram, K. Sawano, Z. Xue, M. Zhang, Z. Di, *ECS transactions*, **93** 73 (2019).
14. C.-M. Lim, Z. Zhao, K. Sumita, K. Toprasertpong, M. Takenaka, and S. Takagi, *IEEE Electron Device Lett.*, **41**, 985 (2020).
15. Y. Moriyama, K. Ikeda, S. Takeuchi, Y. Kamimuta, Y. Nakamura, K. Izunome, A. Sakai, and T. Tezuka, *Appl. Phys. Express*, **7**, 086501 (2014).

16. Y. Nagatomi, S. Tanaka, Y. Nagaoka, K. Yamamoto, D. Wang, and H. Nakashima, *Jpn. J. Appl. Phys.*, **54** 070306 (2015).
17. Y. Nagatomi, T. Tateyama, S. Tanaka, K. Yamamoto, D. Wang, and H. Nakashima, *Semicond. Sci. Technol.*, **32** 035001 (2017).
18. S. Matsuo, T. Fujii, K. Hasebe, K. Takeda, T. Sato, and T. Kakitsuka, *Opt. Express* **22** 12139 (2014).
19. T. Maekura, K. Tanaka, C. Motoyama, R. Yoneda, K. Yamamoto, H. Nakashima, and D. Wang, *Semicond. Sci. Technol.*, **32** 104001 (2017).
20. J. Wasyluk, P. V. Rainey, T. S. Perova, S. J. N. Mitchell, D. W. McNeill, H. S. Gamble, B. M. Armstrong, and R. Hurley, *J. Raman Spectrosc.*, **43**, 448 (2012).
21. H. Haesslein, R. Sielemann, and Christian Zistl, *Phys. Rev. Lett.*, **80**, 2626, (1998).

ECS Transactions, 104 (4) 167-179 (2021)
10.1149/10404.0167ecst ©The Electrochemical Society

Effect of Strain on the Epitaxy of B-Doped $Si_{0.5}Ge_{0.5}$ Source/Drain Layers

G. Rengo[a,b,c], C. Porret[b], A. Y. Hikavyy[b], E. Rosseel[b], M. Ayyad[b], R. J. H. Morris[b], G. Pourtois[b,d], R. Loo[b], and A. Vantomme[a]

[a] Quantum Solid-State Physics, KU Leuven, Celestijnenlaan 200D, 3001 Leuven, Belgium
[b] Imec, Kapeldreef 75, 3001 Leuven, Belgium
[c] FWO - Vlaanderen, Egmontstraat 5, 2000 Brussel, Belgium
[d] Plasmant, University of Antwerp, 2610 Wilrijk-Antwerpen, Belgium

The impact of strain on the growth of *in situ* boron doped $Si_{0.5}Ge_{0.5}$ epitaxial layers is discussed. The lattice strain has been varied by changing the $Si_{0.5}Ge_{0.5}$ thickness and by growing the epitaxial layer on strain relaxed substrates with different Ge concentrations. With decreasing compressive strain, the B incorporation reduces, and the Ge concentration increases. Through density functional theory calculations, the dependence on the applied strain of the energetic cost for boron incorporation into the $Si_{0.5}Ge_{0.5}$ surface was investigated.

Introduction

The progressive downscaling of field-effect transistor (FET) devices is nowadays hindered by the raise in parasitic resistances which makes performance improvements difficult to achieve. In particular, the contact resistance (R_c) at the source/drain (S/D) contacts has become the major component among the parasitics, due to the continuous contact area (A_c) shrinkage (1). To keep this resistance sufficiently low, materials must be engineered to reduce the specific contact resistivity (ρ_c) of the metal/semiconductor stack. In ohmic contacts, ρ_c is linked to the Schottky barrier height ($q\phi_B$) and to the semiconductor carrier concentration (N) via the expression: $\rho_c \propto \exp(q\phi_B/\sqrt{N})$ (2). The need for ultra-low ρ_c therefore initiated renewed efforts to achieve the highest possible active doping concentration in the region of the semiconductor adjacent to the contact.

In pMOS devices, boron doped $Si_{1-x}Ge_x$ is typically used as a S/D material. Due to its larger lattice parameter than silicon, it induces a compressive strain in the channel region that enhances the holes mobility, leading to an overall improvement in the device performances (3). It also enables higher metastable $[B]_{act}$ values than those obtained for pure Si and Ge. The highest B levels reported in literature are close to 10^{21} at./cm^3 and were obtained for Ge concentrations between 50% and 70% (4). However, despite these interesting features, a further increase of the electrically active doping concentration is needed to realize contacts with ρ_c below 1×10^{-9} $\Omega.cm^2$, a value indicated as the milestone for enabling devices beyond the 5 nm node (5).

Several groups have shown that, in nanoscale 3D structures, the misfit biaxial strain in $Si_{1-x}Ge_x$/Si heterostructures is hardly maintained due to the elastic relaxation at the free surfaces (6,7). In addition to a detrimental effect on the channel properties, it also affects

167

the solid solubility of dopants in semiconductors (8). For instance, the compressive strain in $Si_{1-x}Ge_x$ is alleviated by the presence of boron atoms in substitutional position and this elastic energy reduction is reflected in an enhancement of the B solubility (9). Therefore, it is important to understand how the expected strain variations will affect the epitaxial growth processes in future device architectures. Reactions occurring at the growing surface during the chemical vapor deposition (CVD) of Si:B / Si(001) in presence of B_2H_6 have been described in literature (10). However, to the best of the authors' knowledge, a description of the interaction of this precursor with the $Si_{1-x}Ge_x$ surface is still missing.

This paper describes the impact of strain on the growth of *in situ* boron doped $Si_{0.5}Ge_{0.5}$. The study is two-fold: experimental results are first used to describe how the epitaxial growth and the chemical composition of $Si_{0.5}Ge_{0.5}$:B layers vary with the strain state of the growing layer. In the second part, *ab initio* density functional theory (DFT) calculation results are discussed to illustrate how the B_2H_6 decomposition on a $Si_{0.5}Ge_{0.5}$ surface and the B incorporation in the layer are affected by the strain level and strain type (compressive versus tensile).

Experimental details

Epitaxy and characterization methodology

B doped $Si_{0.5}Ge_{0.5}$ epitaxial layers were grown in an ASM Intrepid® production cluster on blanket n-type Si(001) substrates or commercially available $Si_{1-y}Ge_y$(001) strain-relaxed buffers (SRBs) with different Ge contents (25%, 50%, and 70%) (11). The cluster includes two epi reactors and an integrated pre-epi clean module (Previum®). The Si wafer received a high temperature bake (1050°C) in a reduced pressure of H_2 to remove the native oxide. Since this temperature is not compatible with $Si_{1-x}Ge_x$ materials, the SRB substrates underwent a low temperature plasma treatment in the Previum® chamber prior to deposition (12).

All depositions were done by co-flowing conventional precursors such as dichlorosilane ($Si_2Cl_2H_2$), germane (GeH_4), HCl, and diborane (B_2H_6) for the *in situ* B doping. The growth was performed at 550°C and a pressure of 20 Torr.

Surface morphologies were inspected by top-view scanning electron microscopy (SEM) using a KLA-Tencor eDR7100™ electron-beam wafer defect review system. Reciprocal space maps (RSM) were acquired around the asymmetric Si(113) Bragg reflection with a Bruker JVX7300M diffractometer, using a Cu $K_{\alpha1}$ source ($\lambda = 1.5406$ Å) and a two-bounce Ge(220) monochromator. From the RSM, the degree of strain relaxation and the Ge concentration of the epilayers was extracted using a simulation software from Bruker based on the Bragg's law and the modified Vegard's law (13) combined with the Poisson ratio to determine the out of plane lattice constant of the materials. The boron chemical concentrations ($[B]_{chem}$) in the epilayers were measured by secondary ion mass spectroscopy (SIMS) using a CAMECA® SC-Ultra system with an O_2^+ primary beam at an impact energy of 350 eV. In order to quantify the B atomic concentration, five boron implants of known fluence in $Si_{1-x}Ge_x$ with different but known stoichiometry were measured, thus allowing the $[B]_{chem}$ quantification through polynomial fitting. Thicknesses of the layers were also extracted through the crater depth measurement. The Ge concentrations in the samples were measured by Rutherford backscattering spectroscopy (RBS) in random geometry. The incident beam of He^+ ions was produced with a 5SHD-2 linear Pelletron Accelerator manufactured by NEC. An incident energy of 1.57 MeV was used, and the backscattered ions were collected at a glancing-exit angle.

Density functional theory calculations

Ab initio DFT calculations were performed using the Gaussian and Plane-Wave method (GPW) (14) implemented in the Quickstep module of the CP2K software package (15). Simulations were run using the generalized gradient approximation with a modified Perdew-Burke-Ernzerhof exchange correlation functional (PBEsol) (16)(17). The DZVP-MOLOPT-GTH (valence double zeta plus polarization, molecularly optimized) basis sets (18), ideal for calculating properties in gas and condensed phase, were used to describe the electronic density of the atoms present in the system and combined with the pseudopotentials derived by Goedecker-Teter-Hutter (GTH) (19). The cut-offs used for the real space integration of the electronic densities and the Gaussian functions are 350 and 30 Ry, respectively. To sample the Brillouin zone, a $2\times2\times1$ Monkhorst-Pack integration grid is used (20). The electronic temperature has been set to 300 K. For the boron adsorption and incorporation simulations, a $Si_{0.5}Ge_{0.5}(001)$ slab model with eight monoatomic layers and 16 atoms per layer has been developed, in which the Germanium atoms have been distributed randomly. The surface is 2x1 reconstructed and passivated by a hydrogen monolayer. Periodic boundary conditions have been applied to the supercell. In the z-direction the slab is separated from its nearest replica by a 29 Å thick vacuum region. Molecules included in the calculations, namely B_2H_6, BH_3, and H_2, underwent geometry optimization in simulation boxes large enough to avoid interaction with their nearest images. Van der Waals correction schemes (Grimme D3 (21)) were tested on some configurations to verify the impact on calculated enthalpies of formation. Since the energy values were not significantly affected by the corrections, they have not been applied to the calculations here reported.

Results and discussion

Impact of starting template on $Si_{0.5}Ge_{0.5}$ growth

In reference (22), we describe how the incorporation of B and Ge during the $Si_{1-x}Ge_x$ epitaxy on Si substrates is affected by strain relaxation of the growing layer. As an example, Fig. 1a shows the SIMS spectrum as measured for a partially relaxed boron doped $Si_{0.5}Ge_{0.5}$ layer grown on a Si substrate. In the lower part of the layer (33 − 55 nm) both the Ge concentration (x_{Ge}) and $[B]_{chem}$ are constant, except for a B-overshoot at the interface with the substrate, which is caused by an (avoidable) instability in gas pressure when the growth is started. However, in the top 33 nm of the layer, x_{Ge} and $[B]_{chem}$ are not constant. x_{Ge} gradually increases and $[B]_{chem}$ decreases towards the surface. By changing the epi-layer thickness, it has been demonstrated that the variation in Ge and B incorporation coincides with the initiation of strain relaxation during $Si_{1-x}Ge_x$ growth (18). In the first ~2 nm, the quantification of both B and Ge is hindered by the near-surface transient distortion (23). This part is consequently disregarded in the following discussions.

The variation in $[B]_{chem}$ as a function of layer thickness, i.e., the degree of strain relaxation is also reflected in the electrical properties of $Si_{1-x}Ge_x$:B thin films and in the contact properties of $Si_{1-x}Ge_x$:B/Ti stacks (18). Figure 1b shows the resistivity (ρ) and the contact resistivity (ρ_c) as a function of layer thickness, in other words for layers with a different degree of strain-relaxation. The lowest ρ and ρ_c values are obtained for a fully strained 23 nm thick $Si_{0.5}Ge_{0.5}$ layer. Once the metastable critical thickness is exceeded (~25 nm in case of $Si_{0.5}Ge_{0.5}$), both ρ and ρ_c increase with thickness. For the thinnest layer, surface scattering effects play an important role, causing an increase in both ρ and ρ_c. For $Si_{1-x}Ge_x$ layers with a lower Ge concentration (not shown here), the critical thickness is higher and ρ and ρ_c show a plateau as long as the layer thickness does not exceed the metastable critical thickness for strain relaxation (18).

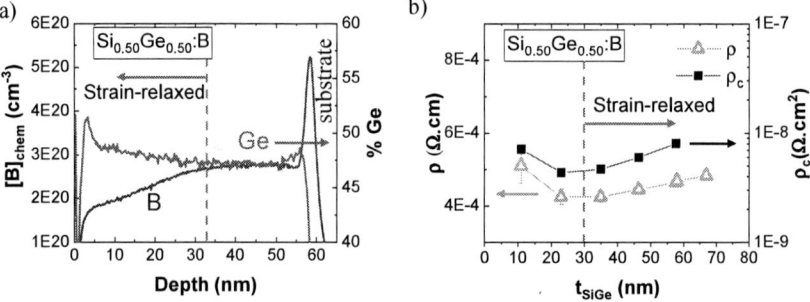

Figure 1. a) B and Ge concentration profiles as measured by SIMS in a partially relaxed $Si_{0.5}Ge_{0.5}$:B epi-layer (18). The region of the layer between the interface with the substrate and the vertical dashed line was grown while being fully strained. The part of the layer on the left side of the red dashed line was deposited after the initiation of strain relaxation. b) ρ and ρ_c as measured for $Si_{0.5}Ge_{0.5}$:B / Ti contacts as a function of the $Si_{0.5}Ge_{0.5}$:B layer thickness (18). The red dashed line indicates the thickness at which strain-relaxation starts to occur.

A similar effect of strain relaxation on material composition has been reported for GeSn, with a spontaneous-relaxation-enhanced Sn incorporation (24–26). An explanation, based on thermodynamic considerations, has been proposed, which links this behaviour to the hindrance of Sn incorporation by the compressive strain present in the material. However, besides the effects of the strain induced by lattice mismatch, local strain fields created by misfit and threading dislocations need to be considered. These local strain variations alter the surface morphology, eventually affecting the material composition during the growth. In this work, this latter effect has been ruled out. The magnitude and sign of the strain have not been varied by inducing strain relaxation in the layers, but by growing $Si_{0.5}Ge_{0.5}$:B epilayers on $Si_{1-y}Ge_y$ SRBs with different compositions (y = 0, 0.25, 0.5, or 0.70). Figure 2 reports the names of the samples, the corresponding stack description, and the strain statuses of the epilayers. By tuning the lattice mismatch between the SRB and the epilayer, it is possible to induce compressive, no-, or tensile strain in the epilayer. The very low threading dislocation density (TDD) in the stack (11) allows to attribute differences in the properties of the grown materials exclusively to the different strain present. The $Si_{0.5}Ge_{0.5}$:B epilayers have a nominal thickness of 60 nm and a $[B]_{chem}$ of 3×10^{20} cm^{-3}.

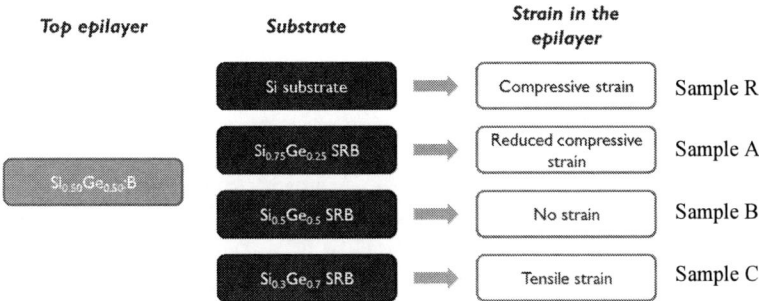

Figure 2. Strain status in the grown epilayers as defined by the mismatch with the underlying virtual substrate.

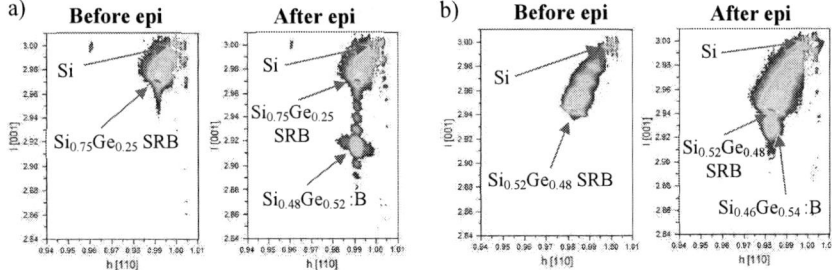

Figure 3. Reciprocal space maps acquired around the Si(113) Bragg reflection for nominal $Si_{0.5}Ge_{0.5}$:B epilayers grown on $Si_{1-y}Ge_y$ SRBs with a) y = 0.25 and b) y = 0.5, acquired before and after epitaxial deposition.

Figure 3 shows the RSMs around the (113) Bragg reflection of Si as obtained for $Si_{0.5}Ge_{0.5}$:B layers grown on SRBs with y = 0.25, and y = 0.50, respectively. Labels indicate the diffraction peaks of the Si substrate, the $Si_{1-y}Ge_y$ SRB, and the $Si_{1-x}Ge_x$ epilayer. The step-graded Ge profile of the SRB signature is clearly recognized (11). The peak with the highest intensity corresponds to the top layer of the SRB. The SRBs are nearly fully relaxed. In Figure 3a - after epi, as expected, the epilayer peak is located below the $Si_{0.75}Ge_{0.25}$ diffraction peak, confirming its larger out of plane lattice constant. In Figure 3b, the epilayer peak was ideally supposed to overlap with that of the SRB as the nominal x_{Ge} was 0.50 for both. Instead, their position differ slightly. A Ge fraction of ~ 0.48 has been extracted for the SRB top layer, and ~ 0.54 for the epilayer. It is also noticed that the extracted x_{Ge} in the grown layer progressively increases with increasing y_{Ge} in the SRB. For the examples shown in Figure 3, the peaks assigned to the $Si_{0.5}Ge_{0.5}$:B epilayers appear at the same h coordinates as the top of the underlying SRBs, meaning that their in-plane lattice parameters match. These epilayers are therefore fully strained with respect to the underlying SRB. The layer grown on Si has, instead, a degree of relaxation of 25%. However, before reaching the critical thickness, the layer had grown pseudomorphically, while containing the highest compressive strain among all the samples here reported. The region of the layer adjacent to the substrate is therefore relevant for comparison with the layers grown on SRBs. The extracted Ge concentrations and degrees of relaxation for the virtual substrates and for the epilayers are reported in TABLE I. The relaxation is indicated with respect to Si for the SRB top layers, and with respect to the SRB for the epilayers.

TABLE I. Ge concentrations and degrees of strain relaxation of the virtual substrates and the epilayers as extracted from RSMs.

Sample	SRB			Epilayer		
	Nominal material	y_{Ge}	Degree of Relaxation (%)	Nominal material	x_{Ge}	Degree of Relaxation [*wrt* SRB]
R	Si	0	100	$Si_{0.5}Ge_{0.5}$:B	0.48	25%
A	$Si_{0.75}Ge_{0.25}$	0.25	94	$Si_{0.5}Ge_{0.5}$:B	0.52	~0%
B	$Si_{0.5}Ge_{0.5}$	0.48	97	$Si_{0.5}Ge_{0.5}$:B	0.54	~0%
C	$Si_{0.3}Ge_{0.7}$	0.66	99	$Si_{0.5}Ge_{0.5}$:B	NA	NA

Figure 4a compares the B concentration depth profiles as extracted by SIMS for the B-doped $Si_{1-x}Ge_x$ epitaxial layers grown on Si and on the different SRBs. The B incorporation during epitaxial growth is clearly affected by the magnitude and the type of the biaxial strain. The highest B concentrations have been measured for the compressively strained $Si_{0.5}Ge_{0.5}$ layers. It is, however, remarkable that the $Si_{0.5}Ge_{0.5}$ layer grown on the SRB with 25% Ge contains a similar boron level to the layer grown on the Si substrate. This is also the case just above the substrate/epitaxial interface, i.e. when the $Si_{0.5}Ge_{0.5}$ layer grown on Si was still fully strained. As expected from 25 nm above the interface on, the boron incorporation decreases due to the onset of strain relaxation. In the nominally unstrained layer, the [B]$_{chem}$ is significantly lower. The $Si_{0.5}Ge_{0.5}$ grown on the $Si_{0.3}Ge_{0.7}$ SRB has the lowest [B]$_{chem}$ of ~ 2 x 10^{19} cm^{-3}. This indicates that the tensile strain is detrimental for B incorporation.

The Ge concentration in the epilayer, as extracted from RBS, is also significantly affected by the starting template. It, however, shows an opposite trend compared to [B]$_{chem}$, with a monotonic increase in x_{Ge} as the compressive strain is reduced and tensile strain introduced. Results from this assessment are summarized in TABLE II. The $Si_{1-x}Ge_x$ layer grown on the $Si_{0.5}Ge_{0.5}$ SRB for instance presents a Ge fraction being 8% higher than that of the epilayer grown on Si.

Finally, the growth rates of the epilayers are also significantly impacted. Since B doping is known to enhance the growth rate in the CVD of Si (27), the reduction in boron incorporation with reducing the compressive strain may explain the significant decrease in growth rate from sample A to B. As different starting substrates could affect the surface temperature during growth, resulting in the differences evidenced above, surface temperatures were monitored used a pyrometer during the process. Temperature differences were found to be very limited, in the order of 1-2°C, which rules out a significant impact on the epitaxial growth process.

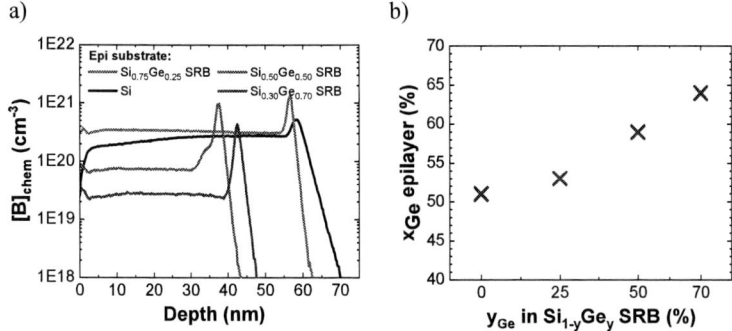

Figure 4.a) SIMS boron concentration profiles as measured for the epilayers of the samples. b) Epilayers Ge concentrations as measured with RBS.

TABLE II. Layer thicknesses and $[B]_{chem}$ as extracted from SIMS, and Ge concentrations measured by RBS.

Sample	Thickness (nm)	Growth rate (nm/min)	$[B]_{chem}$ (cm^{-3})	x_{Ge}
R	59	6.8	2.7×10^{20}	0.51
A	56	6.7	3.0×10^{20}	0.53
B	37	4.4	7.3×10^{19}	0.59
C	43	5.2	2.7×10^{19}	0.64

DFT modelling of B_2H_6 decomposition and B incorporation at the $Si_{0.5}Ge_{0.5}$ surface

Atomistic DFT calculations were used to investigate the adsorption and decomposition of diborane on a (001) oriented $Si_{0.5}Ge_{0.5}$ surface and to predict how these processes are affected by different strain levels. A schematic of the typical slab model is shown in Figure 5a. The models are built starting from an eight atomic layers thick Si(001) slab. The exposed surfaces are 2x1 reconstructed and contain 16 Si atoms per layer, framed in two rows of four dimers each. Both the bottom and top surfaces are fully H-passivated. The interaction between the top surface and the B_2H_6 molecule is studied, while the bottom surface remains untouched. After atomic positions and cell optimizations, in-plane supercell dimensions of 15.41 Å and 15.16 Å are obtained for x and y, respectively. The Si atoms in the slab are then randomly replaced with Ge atoms, with a probability of 50%. A $Si_{0.5}Ge_{0.5}$(001) slab is obtained, with the Si or Ge atoms placed on the sites corresponding to the pure Si lattice. Next, the geometry is optimized again using four different constraint conditions: (i) the atomic positions and lattice parameters are optimized simultaneously to obtain a $Si_{0.5}Ge_{0.5}$ slab without any strain. The relaxed cell expands to 15.80 Å and 15.54 Å in the x and y directions, respectively. (ii) The lattice constants are fixed to match those of the Si slab. The geometry is optimized (atomic positions but not the simulation cell dimensions) and the atomic coordinates of the system expand in the z-direction into the vacuum region. A slab replicating pseudomorphic $Si_{0.5}Ge_{0.5}$(001) is obtained. This system corresponds to the fully-compressively-strained case. For systems (iii) and (iv), two fixed sets of x and y cell dimensions have been chosen, with intermediate values between those of the fully relaxed (i) and the pseudomorphic (ii) cases, after which the atomic positions have been optimized. The obtained strain statuses correspond to 33% and 66% relaxed $Si_{0.5}Ge_{0.5}$ on

Si. The stress tensors for the four systems are calculated. The typical tensor structure corresponds to that of a biaxially strained system, with comparable σ_{xx} and σ_{yy} components and neglectable stress values in all the other components. Details of the four slab models used are summarized in TABLE III.

TABLE III. Strain conditions applied to the slab model

Strain condition	(x,y) cell dimensions (Å^2)	σ_{xx} (GPa)	σ_{yy} (GPa)
Fully-compressively-strained	15.41 x 15.16	4.72	4.03
33% relaxed	15.54 x 15.29	3.06	2.58
66% relaxed	15.68 x 15.43	1.36	1.09
Unstrained	15.80 x 15.54	0.00	0.00

The decomposition pathway of a B_2H_6 molecule on the $Si_{1-x}Ge_x(001)$ surface is considered to proceed in three steps, namely (i) the chemisorption of the B_2H_6 molecule on the surface, (ii) surface reaction, and incorporation in the lattice by either replacing a Ge (iii-a) atom or a Si (iii-b) atom (Figure 5b-e). The surface chemical reactions corresponding to the different steps are indicated in the following. The symbol (s) indicates that the compound is a solid. It is in the gaseous phase otherwise.

$$Si_{0.5}Ge_{0.5}(s) + B_2H_6 \rightarrow Si_{0.5}Ge_{0.5} - B_2H_5(s) + \frac{1}{2}H_2 \qquad \text{(i)}$$

$$Si_{0.5}Ge_{0.5}(s) + B_2H_6 \rightarrow Si_{0.5}Ge_{0.5} - BH_2(s) + BH_3 + \frac{1}{2}H_2 \qquad \text{(ii)}$$

$$Si_{0.5}Ge_{0.5}(s) + B_2H_6 \rightarrow Si_{0.5}Ge_{0.5(-1Ge\ at.)}:B - GeH_2(s) + BH_3 + \frac{1}{2}H_2 \qquad \text{(iii-a)}$$

$$Si_{0.5}Ge_{0.5}(s) + B_2H_6 \rightarrow Si_{0.5(-1Si\ at.)}Ge_{0.5}:B - SiH_2(s) + BH_3 + \frac{1}{2}H_2 \qquad \text{(iii-b)}$$

The formation enthalpy (ΔH_f) for each step with respect to the initial system is given by:

$$\Delta H_{f,step} = (E_{ads} + E_{bypr}) - (E_{slab} + E_{B_2H_6}) \qquad [1]$$

where E_{ads} is the DFT-calculated energy for the SiGe slab with the B_xH_y, GeH_2, or SiH_2 chemisorbed group, E_{bypr} is the energy of the gaseous by-product(s), E_{slab} is the energy of the pristine $Si_{0.5}Ge_{0.5}(001)$ surface slab, and E_{B2H6} is the energy of the gaseous B_2H_6 molecule. This analysis does not consider thermal effects, such as the entropy variation, but provides indications of the most stable equilibrium configurations in the systems slab + adsorbates. However, the epitaxial growth is an out-of-equilibrium process, which cannot be fully described in this way. Nevertheless, the results obtained from DFT modelling still provide information about the driving forces for the layer deposition.

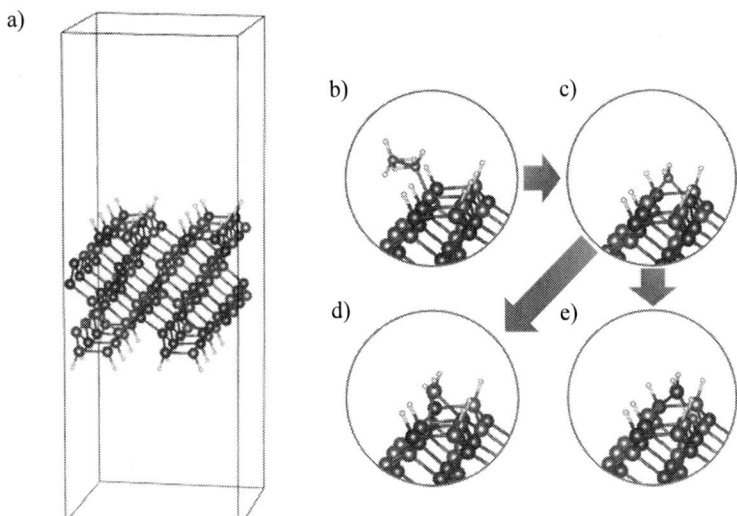

Figure 5.a) $Si_{0.5}Ge_{0.5}(001)$-2x1 surface slab model used in DFT calculations. b) "Adsorption" step. c) "Reaction" step. d) "Incorporation (Ge)" step. e) Incorporation (Si)" step.

Figure 6 shows the formation enthalpies (ΔH_f) calculated for the different steps of the B_2H_6 reaction pathway the $Si_{0.5}Ge_{0.5}$ surface and for applying different magnitudes of strain to the slab. Because the chosen initial system is very stable with a fully hydrogen passivated reconstructed surface, ΔH_f has positive values for all the configurations. In reality, the deposition process proceeds at elevated temperatures and the hydrogen passivation is partially lost thus creating potential adsorption sites for the B_2H_6 molecules. For the adsorption and reaction steps (represented in Figure 5b-c) the calculated ΔH_f is hardly affected by the amount of applied strain. The incorporation of a B atom into the surface is simulated by starting from the "Reaction" configuration and swapping the atomic positions of B and either a Ge atom in the surface dimer (Figure 5d), or a Si atom (Figure 5e) as proposed in (28) for the dissociation of phosphine on a Si(001) surface. Two different configurations are obtained, where the boron atom replaces one of the lattice atoms in a surface dimer and the removed Ge or Si atom is placed on top of the surface as an adatom. In both cases the calculated formation enthalpies vary with the magnitude of compressive strain present in the surface slab. From this result, one can expect that the epitaxy *in situ* boron doped epitaxial $Si_{1-x}Ge_x$ growth is affected by the stain state of the growing layer. A higher incorporation probability of the B atoms is expected during the growth of compressively strained layers with respect to unstrained layers.

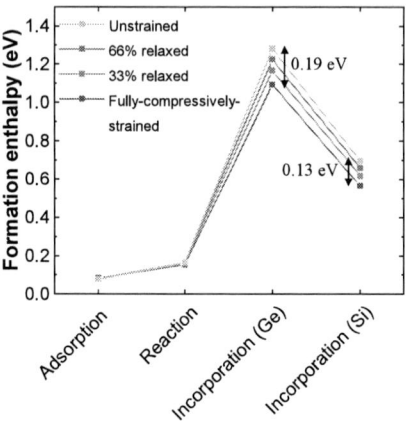

Figure 6. Formation enthalpies calculated for intermediate steps of the B_2H_6 decomposition and B incorporation in $Si_{0.5}Ge_{0.5}(001)$ with different strain conditions.

In Figure 7a, we report the evolution of the x-component of the stress in the slab at the various steps of the B_2H_6 decomposition pathway and for different strain states. Positive values correspond to compressive stress, negative to tensile. As expected, the chemisorbed B_xH_y groups have a limited impact on the lattice strain, while the B incorporated into the surface row of the atomic lattice causes a release of the compressive strain. The strain release is larger when a Ge atom is replaced by B, rather than a Si atom. The reduction in stress is highest for the fully compressively strained lattice and decreases with decreasing strain in the starting slab (Fig. 7b). This supports the experimentally observed tendency of B to be incorporated when a compressive strain is present in the growing layer and the increase in x_{Ge} when the compressive strain is not present.

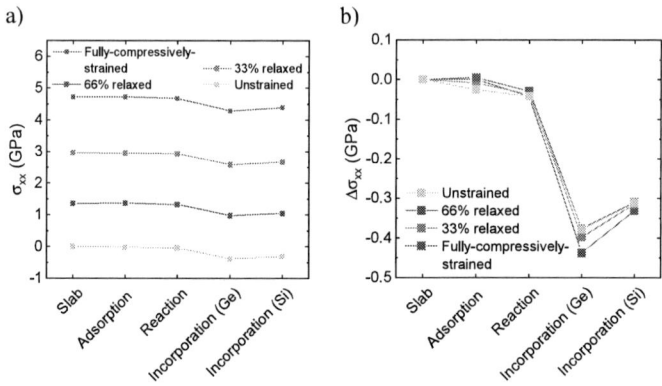

Figure 7. a) x-component of calculated stress in the slabs for the different steps of the considered B_2H_6 decomposition pathway and for different strain conditions. b) Relative

variation in the x-component of the stress with respect to that of the pristine surface slab for the different decomposition steps. The four applied strain conditions are reported.

Summary

We have described how strain affects the epitaxial growth process of *in situ* B doped $Si_{0.5}Ge_{0.5}$. In case of compressively strained $Si_{0.5}Ge_{0.5}$:B growth on Si, B and Ge incorporation during the epitaxial growth are affected by the initiation of layer relaxation, i.e., the B (Ge) incorporation decreases (increases) with increasing degree of strain relaxation. The effect of elastic strain on B and Ge incorporation was also observed when modifying the magnitude and sign of the strain by growing B doped $Si_{0.5}Ge_{0.5}$ layers onto $Si_{1-y}Ge_y$ strain relaxed buffers (SRBs) with different compositions.

Through *ab initio* density functional theory calculations, the mechanisms of B_2H_6 decomposition and the effect of strain on the B incorporation into a $Si_{0.5}Ge_{0.5}(001)$ surface have been simulated while applying four different strain conditions. The energetic cost to incorporate the B atom into the first atomic layer of the surface varies with the strain applied to lattice. The obtained trends in energetic cost vs strain are in line with the experimental observations. The calculation of the stress evolution in the slab model at the different B_2H_6 reaction steps confirms that B incorporation in the lattice reduces the compressive strain caused by the presence of Ge in the lattice. The change in lattice strain by B incorporation increases with increasing compressive strain implemented in the starting model. The other steps of the reaction pathway do not have a significant impact on the strain in the slab. Therefore, the variation in compressive strain is expected to influence the incorporation probability of boron during the epitaxial deposition, while the surface coverage of B_xH_y-groups is expected to be unaffected by the strain status of the layer.

Acknowledgments

G. Rengo acknowledges the Research Foundation of Flanders (FWO) for granting him a PhD fellowship strategic basic research. The imec core CMOS program members, European Commission, local authorities and the imec pilot line are acknowledged for their support. This project has received funding from the ECSEL Joint Undertaking (JU) under grant agreement No 875999. The JU receives support from the European Union's Horizon 2020 research and innovation programme and Netherlands, Belgium, Germany, France, Austria, Hungary, United Kingdom, Romania, Israel. The imec MSP group is acknowledged for useful weekly discussions. We gratefully acknowledge Qiang Zhao from the QSP unit of KU Leuven for his support with the RBS measurements. The $Si_{1-x}Ge_x$ SRB wafers were provided by Siltronic AG.

References

[1] A. V.-Y. Thean, D. Yakimets, T. Huynh Bao, P. Schuddinck, S. Sakhare, M. G. Bardon, A. Sibaja-Hernandez, I. Ciofi, G. Eneman, A. Veloso, J. Ryckaert, P. Raghavan, A. Mercha, A. Mocuta, Z. Tokei, D. Verkest, P. Wambacq, K. De Meyer, and N. Collaert, *2015 IEEE Symp. VLSI Technol.*, pp. T26–T27 (2015).

[2] S. M. Sze and K. K. Ng, *Physics of Semiconductor Devices*. Hoboken, NJ, USA,

2006.

[3] P. Verheyen, N. Collaert, R. Rooyackers, R. Loo, D. Shamiryan, A. De Keersgieter, G. Eneman, F. Leys, A. Dixit, M. Goodwin, Y. S. Yim, M. Caymax, K. De Meyer, P. Absil, M. Jurczak, and S. Biesemans, *2005 IEEE Symp. VLSI Technol.*, pp. 194–195 (2005).

[4] R. Loo, A. Y. Hikavyy, L. Witters, A. Schulze, H. Arimura, D. Cott, J. Mitard, C. Porret, H. Mertens, P. Ryan, J. Wall, K. Matney, M. Wormington, P. Favia, O. Richard, H. Bender, A. Thean, N. Horiguchi, D. Mocuta, and N. Collaert, *ECS J. Solid State Sci. Technol.*, **6**(1), pp. P14–P20 (2017).

[5] C.-N. Ni, K. V. Rao, F. Khaja, S. Sharma, S. Tang, J. J. Chen, K. E. Hollar, N. Breil, X. Li, M. Jin, C. Lazik, J. Lee, H. Maynard, N. Variam, A. J. Mayur, S. Kim, H. Chung, M. Chudzik, R. Hung, N. Yoshida, and N. Kim, *2016 IEEE Symp. VLSI Technol.*, pp. 1–2 (2016).

[6] O. Moutanabbir, M. Reiche, W. Erfurth, F. Naumann, M. Petzold, and U. Gösele, *Appl. Phys. Lett.*, **94**(24), pp. 1–4 (2009).

[7] M. Tomita, D. Kosemura, K. Usuda, and A. Ogura, *ECS Trans.*, **53**(1), pp. 207–214 (2013).

[8] C. Ahn, N. Bennett, S. T. Dunham, and N. E. B. Cowern, *Phys. Rev. B - Condens. Matter Mater. Phys.*, **79**(7), pp. 1–4 (2009).

[9] B. Tillack, P. Zaumseil, G. Morgenstern, D. Krüger, and G. Ritter, *Appl. Phys. Lett.*, **67**, pp. 1143–1144 (1995).

[10] Q. Campbell, J. A. Ivie, E. Bussmann, S. W. Schmucker, A. D. Baczewski, and S. Misra, **125**(1), pp. 481–488 (2021).

[11] G. Kozlowski, O. Fursenko, P. Zaumseil, T. Schroeder, M. Vorderwestner, and P. Storck, *ECS Trans.*, **50**(9), pp. 613–621 (2013).

[12] F. Wang, B. B. Jotheeswaran, J. Tolle, X. Lin, P. Gao, and A. Demos, *Solid State Phenom.*, **282**, pp. 25–30 (2018).

[13] J. P. Dismukes, L. Ekstrom, and R. J. Paff, *J. Phys. Chem.*, **68**(10), pp. 3021–3027 (1964).

[14] G. Lippert, J. Hutter, and M. Parrinello, *Mol. Phys.*, **92**(3), pp. 477–487 (1997).

[15] J. Vandevondele, M. Krack, F. Mohamed, M. Parrinello, T. Chassaing, and J. Hutter, *Comput. Phys. Commun.*, **167**(2), pp. 103–128 (2005).

[16] J. P. Perdew, K. Burke, and M. Ernzerhof, *Phys. Rev. Lett.*, **77**(18), pp. 3865–3868 (1996).

[17] J. P. Perdew, A. Ruzsinszky, G. I. Csonka, O. A. Vydrov, G. E. Scuseria, L. A. Constantin, X. Zhou, and K. Burke, *Phys. Rev. Lett.*, **100**(13), pp. 1–4 (2008).

[18] J. VandeVondele and J. Hutter, *J. Chem. Phys.*, **127**(11), p. 114105 (2007).

[19] S. Goedecker and M. Teter, *Phys. Rev. B - Condens. Matter Mater. Phys.*, **54**(3), pp. 1703–1710 (1996).

[20] H. J. Monkhorst and J. D. Pack, *Phys. Rev. B*, **13**(12), pp. 5188–5192 (1976).

[21] S. Grimme, J. Antony, S. Ehrlich, and H. Krieg, *J. Chem. Phys.*, **132**(15), p. 154104 (2010).

[22] G. Rengo, C. Porret, A. Y. Hikavyy, E. Rosseel, N. Nakazaki, G. Pourtois, A. Vantomme, and R. Loo, *ECS Trans.*, **98**(5), pp. 27–36 (2020).

[23] W. Vandervorst, T. Janssens, B. Brijs, T. Conard, C. Huyghebaert, J. Frühauf, A. Bergmaier, G. Dollinger, T. Buyuklimanli, J. A. VandenBerg, and K. Kimura, *Appl. Surf. Sci.*, **231–232**, pp. 618–631 (2004).

[24] J. Margetis, S. Al-Kabi, W. Du, W. Dou, Y. Zhou, T. Pham, P. Grant, S. Ghetmiri, A. Mosleh, B. Li, J. Liu, G. Sun, R. Soref, J. Tolle, M. Mortazavi, and S. Q. Yu,

ACS Photonics, **5**(3), pp. 827–833 (2018).

[25] W. Dou, M. Benamara, A. Mosleh, J. Margetis, P. Grant, Y. Zhou, S. Al-Kabi, W. Du, J. Tolle, B. Li, M. Mortazavi, and S. Q. Yu, *Sci. Rep.*, **8**(1), pp. 1–11 (2018).

[26] S. Assali, J. Nicolas, and O. Moutanabbir, *J. Appl. Phys.*, **125**(2), (2019).

[27] D. Grützmacher, *J. Cryst. Growth*, **182**(1–2), pp. 53–59 (1997).

[28] H. F. Wilson, O. Warschkow, N. A. Marks, S. R. Schofield, N. J. Curson, P. V. Smith, M. W. Radny, D. R. McKenzie, and M. Y. Simmons, *Phys. Rev. Lett.*, **93**(22), pp. 2–5 (2004).

Chapter 7

Novel Materials and Characterization 2

ECS Transactions, 104 (4) 183-189 (2021)
10.1149/10404.0183ecst ©The Electrochemical Society

(Invited) Thermoelectric Properties of Tin-Incorporated Group-IV Thin Films

Masashi Kurosawa[a], and Osamu Nakatsuka[a,b]

[a] Graduate School of Engineering, Nagoya University, Nagoya 464-8603, Japan
[b] Institute of Materials and Systems for Sustainability, Nagoya University, Nagoya 464-8601, Japan

We investigate a new application of germanium tin ($Ge_{1-x}Sn_x$) binary alloy thin films to realize energy harvesting of low-grade heat to electricity, i.e., thin-film thermoelectric generator (TEG). To clarify the potential of $Ge_{1-x}Sn_x$ for the TEG application experimentally, it needs to choose high-resistivity wafers as the substrate for the $Ge_{1-x}Sn_x$ growth to isolate electrically from the substrates. Specifically, this paper conducts crystal growths of $Ge_{1-x}Sn_x$ on FZ-Si, semi-insulating GaAs, and InP substrates. The impacts of Sn content and crystallographic tilt in the $Ge_{1-x}Sn_x$ on the thermal conductivity will be discussed experimentally and theoretically. We also show the scaling merit of the device sizes in the power density.

Introduction

Integration of a thin-film energy-harvesting device in a silicon (Si) chip strongly desires to realize a standalone sensing network system called the internet-of-things (IoT) society. The thermoelectric generator (TEG), which could convert waste heat into electricity, is one option. Considering many IoT sensors will be distributed worldwide, it is desirable to manufacture them using the same technology as next-generation Si-based integrated circuits, including the material. Tin-incorporated group-IV materials, such as germanium tin ($Ge_{1-x}Sn_x$) and silicon tin ($Si_{1-x}Sn_x$), theoretically possess a lower thermal conductivity (1) than those of Ge and Si, respectively, owing to the mass-difference scattering of phonons. Thermoelectric efficiency is inversely proportional to thermal conductivity; hence, the tin-incorporated group-IV thin films are one of the attractive TEG materials. The low thermal conductivity also helps ensure temperature differences within small TEG devices.

In our previous study on molecular beam epitaxy (MBE)-grown $Ge_{1-x}Sn_x$ layers (2), it was found that the thermal conductivity was decreased (to 2.6 $Wm^{-1}K^{-1}$) with increasing the Sn content (to 12 %). Recently, Spirito et al. reported a low thermal conductivity of 4 $Wm^{-1}K^{-1}$ for the chemical vapor deposition (CVD)-grown $Ge_{1-x}Sn_x$ layers with the Sn content of 14% (3). These results are quite informative for the TEG applications; however, the thermoelectric properties have not been clarified experimentally, compared with the polycrystalline films such as $Ge_{1-x}Sn_x$ (4-6), $Si_{1-x}Sn_x$ (7), and $Si_{1-x-y}Ge_xSn_y$ (8,9). Against the background, we have discussed the thermoelectric properties of the $Ge_{1-x}Sn_x$ layers grown on high-resistivity substrates. Due to the limited space of the paper, we will show here two topics of thermal conductivity and TEG performances. The thermal conductivity

183

of $Ge_{0.95}Sn_{0.05}$ was reduced down to 2.5 $Wm^{-1}K^{-1}$ (10,11) by introducing the stacking faults due to microscopic tilt in the crystals, which value is 4% of bulk-Ge (60 $Wm^{-1}K^{-1}$ (12)).

Experimental Methods

Crystal growth

We chose high-resistivity wafers (at least exceeding 10^4 Ωcm) as the substrate for the $Ge_{1-x}Sn_x$ growth to isolate electrically from the substrates, specifically, semi-insulating GaAs(001), InP(001), and n-type FZ-Si(001) wafers with a 3-inch diameter. First, the substrates were cleaned by conventional wet-chemical treatment and then heat treatment in an ultra-high vacuum chamber with a base pressure of 10^{-7}–10^{-8} Pa. The details of the treatments of Si, GaAs, and InP wafers were described elsewhere (11,13,14). Next, a 100-nm-thick (or 150-nm-thick) p- or n-type $Ge_{1-x}Sn_x$ ($x = 0$–0.12) layer was grown on the substrates using Knudsen cells. The p- or n-type dopant was chosen as gallium (Ga) or antimony (Sb), respectively. The Ga and Sb concentrations were ranging from 10^{18} to 10^{20} cm^{-3} and 10^{20} cm^{-3}, respectively.

We used XRD two-dimensional reciprocal space mapping (XRD-2DRSM) to identify the crystallographic structures of the grown layers, commercial equipment (Netzsch, SBA 458 Nemesis) to determine the Seebeck coefficient, Hall effect measurement system (Toyo Corp., ResiTest8300) with the van der Pauw to identify the electrical properties, and finally pico-second thermoreflectance (PicoTherm, PicoTR) or 2ω method (Advance Riko, TCN-2ω) to analyze the thermal conductivity of the $Ge_{1-x}Sn_x$ in the thickness direction. All of the measurements were carried out at room temperature (RT).

Theoretical calculations of thermal conductivity

To calculate the thermal conductivity κ, we used a model based on the Boltzmann transport equation proposed by Callaway and Baeyer (15). The κ is given by

$$\kappa = \frac{k_B}{2\pi^2 v^2} \int_0^{\omega_{max}} \left(\frac{\hbar\omega}{k_B T}\right)^2 \frac{\exp(\hbar\omega/k_B T)}{[\exp(\hbar\omega/k_B T) - 1]^2} \omega^2 \lambda \, d\omega, \qquad [1]$$

where k_B is the Boltzmann constant, T is the absolute temperature, $\hbar$ is the reduced Planck constant, and ω is the phonon frequency. The v is the average sound velocity, given by $v^{-2} = \frac{1}{3}(v_L^{-2} + 2v_T^{-2})$, where v_L is the longitudinal sound velocity, and v_T is the transverse sound velocity. The phonon scattering mechanisms considered in this study are the Sn-induced scattering, umklapp scattering, and domain boundary scattering. Their mean free path (λ_{Sn}, λ_U, and λ_{DB}, respectively) are combined using Matthiessen rule: $\lambda^{-1} = \lambda_{Sn}^{-1} + \lambda_U^{-1} + \lambda_{DB}^{-1}$. These scattering mechanisms can be written as $\lambda_{Sn}^{-1} = A\omega^4/v_s$ (16), $\lambda_U^{-1} = B\omega^2 T \exp(-\theta/3T)/v_s$ (17), and $\lambda_{DB}^{-1} = 1/d$ (18). The scattering parameters are expressed by $A = \frac{V}{4\pi v^2} x(1-x)\left(\frac{\Delta M}{M}\right)$ and $B = \hbar\gamma^2/Mv^2\theta$. In these equations, θ is the Debye temperature expressed by $\theta = \hbar\omega_{max}/k_B$, d is the average domain size, V is the volume per atom, x is the Sn content, M is the average atomic weight, ΔM is the difference of the atomic weight between Sn and Ge, γ is Gruneisen constant, which is chosen as 2 in this study.

Results and Discussions

Domain-size-dependence on thermoelectric properties

This section will discuss the influence of the in-plane correlation length $<L>$ caused by the microscopic tilt in $Ge_{1-x}Sn_x$ crystals on the thermal and electrical conductivity. The details of the estimated method of $<L>$ were described elsewhere (10,11). Figure 1 shows the thermal conductivity κ obtained from the $Ge_{1-x}Sn_x$ layers at various growth conditions: substrate (Si or Ge), dopant (without or Sb), and Sn content (0−12%), where the fill color in the symbols correspond to $<L>$. The calculated values using a model based on the Boltzmann transport equation for different x and the grain size d are also given for comparison as solid curves. It is confirmed that the κ decreases with increasing the x for the calculated results. The calculated curves are gradually shifted to smaller values with falling d from 100 to 1 nm. The trend was also confirmed for the experimental results. It is found that $<L>$ of the $Ge_{1-x}Sn_x$ layers grown on the Ge substrates (diamond symbols) is the largest in these samples, resulting in the highest κ. With decreasing $<L>$ from ~10 to ~2 nm, the κ is gradually shifted to smaller values as with the calculated results. Specifically, the κ of Sb-doped $Ge_{0.95}Sn_{0.05}$ on Si (circle symbol) was reduced to 1.4 $Wm^{-1}K^{-1}$ by introducing the many stacking faults due to the lattice mismatch with the underlying substrate. The value (1.4 $Wm^{-1}K^{-1}$) is 4% of bulk-Ge (60 $Wm^{-1}K^{-1}$ (12)).

We recently found that the stacking faults did not affect hole conduction much in a range of high hole concentrations (not shown here). The power factors for the p- and n-type $Ge_{1-x}Sn_x$ layers grown on GaAs were ~10 and ~30 $\mu Wcm^{-1}K^{-2}$ at RT, respectively. The power factor for the n-type layers is comparable with the counterpart of the n-type BiTe-based layers (~25 $\mu Wcm^{-1}K^{-2}$ at RT (19)). The details will be published elsewhere.

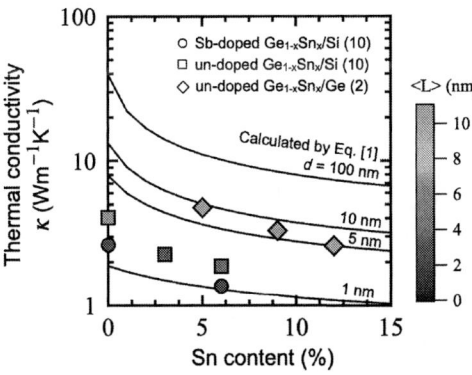

Figure 1. Thermal conductivity κ versus in-plane correlation length $<L>$ for undoped and Sb-doped $Ge_{1-x}Sn_x$ layers grown on various substrates of Si (10) or Ge (2). Solid curves are calculated values using Eq. [1].

Fabrication of uni-leg GeSn thermoelectric generators

It is well known that miniaturization according to the scaling rule is advantageous in terms of power consumption reduction and switching speed improvement in transistors, but the effect of miniaturization in TEG was unclear. Recently, Watanabe *et al.* have proposed the guideline to enhance the TEG power density generated from a unit area in cavity-free Si nanowire TEGs (20). They showed the power density could be enhanced up to 12 μWcm^{-2} by shortening the length of the Si nanowires (21). In this section, $Ge_{1-x}Sn_x$ epitaxial layers were patterned into islands of different sizes, and their thermoelectric power densities were evaluated. The measurement setup was described elsewhere (5,6). The physical properties before the pattering are shown in Table I.

TABLE I. Summary of physical properties of Ga-doped $Ge_{1-x}Sn_x$ layers grown on the semi-insulating GaAs substrates. These properties were analyzed at RT before the device fabrication.

Properties	Analyzed values
Sn content (%)	3
Thickness (nm)	160
Hall hole concentration p (cm^{-3})	3.8×10^{19}
Electrical conductivity σ (S/cm)	583
Seebeck coefficient S $(\mu V/K)$	132

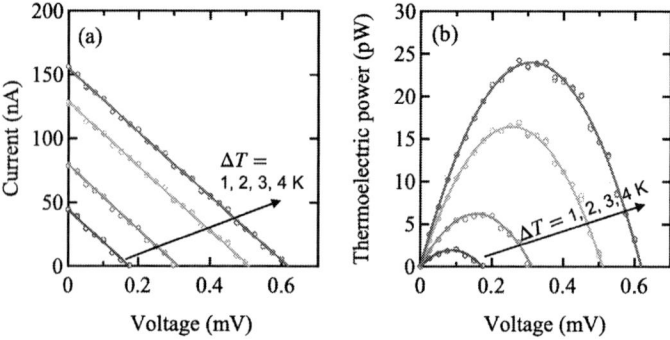

Figure 2. Thermoelectric (a) current-voltage and (b) power-voltage properties of the p-type $Ge_{1-x}Sn_x$ uni-leg TEGs obtained under constant temperature differences ($\Delta T = 1, 2, 3,$ and 4 K). The measurement temperature was RT.

We confirmed that the Seebeck coefficient and the electrical conductivity do not change after the island patterning (not shown here). Figure 2(a) and 2(b) shows the typical thermoelectric current-voltage and power-voltage properties for the sample with a width of 150 μm and a length of 5000 μm, respectively. The thermoelectric power increases with increasing temperature difference ΔT. The maximum power P_{max} was 24 pW at a temperature difference of 4 K. Compared with in-plane π-type TEG in polycrystalline $Ge_{1-x}Sn_x$ layers ($P_{max} \approx 6$ pW at $\Delta T = 4$ K, device area: 700 μm×5000 μm×5 unit) (6), the power generation of the TEG fabricated in this study is about four times higher than that of the previous research, even though the area is 3/700. This can be attributed to the higher electrical conductivity (583 Scm^{-1}) than the previous study (p-type: 223 Scm^{-1}, n-type: 119 Scm^{-1}) and lower contact resistance than at the Al/n-Ge$_{1-x}$Sn$_x$ interface.

Finally, we summarize the change in the maximum power generation per unit area (power density) as a function of ΔT in Fig. 3. The dashed line is the power density when the contact resistance between the Al electrode and the Ge$_{1-x}$Sn$_x$ layers could be ignored. The power densities of thin-film TEG in π-type polycrystalline Ge$_{1-x}$Sn$_x$ (6) is also shown for comparison. It can be seen that the power density of the Ge$_{1-x}$Sn$_x$ uni-leg TEGs improves as the device length L becomes smaller. However, as the ratio of the contact resistance to the resistance of the Ge$_{1-x}$Sn$_x$ itself becomes larger by shortening the device length L, the density drops significantly from the ideal power density (dashed lines). This indicates that reducing the contact resistance is an issue for improving the device performance, as in the case of metal-oxide-semiconductor field-effect transistors. We believe that this issue will be solved by using the Si-based group-IV semiconductor technology, which has been actively discussed in the field of metal contact with Ge (22-25).

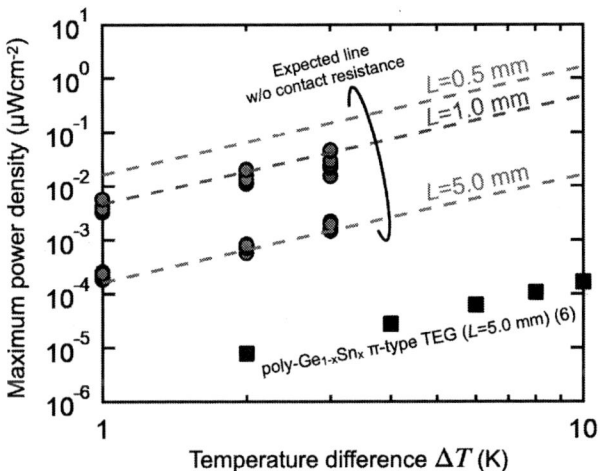

Figure 3. Thermoelectric power density at RT as a function of ΔT. The power density was determined by P_{max}/WL, where P_{max} is the maximum thermoelectric power, W is the device width (3, 5, 50, 150, and 500 μm), and L is the device length (500, 1000, and 5000 μm).

Summary

We have shown here that the potential of epitaxial Ge$_{1-x}$Sn$_x$ layers for thin-film thermoelectric applications operated around room temperature and discussed the scaling merit. Group-IV semiconductors (Ge, Si$_{1-x}$Ge$_x$, Ge$_{1-x}$Sn$_x$) on insulator structure on Si substrate were realized using layer transfer technology proposed by Maeda *et al.* (26,27). Hence, we are expecting that the integration of Ge$_{1-x}$Sn$_x$ TEGs will become possible by using the technique. The experimental results shown in this paper are informative for realizing a thin-film energy-harvesting device in Si chips essential for a standalone sensing network system.

Acknowledgments

The authors thank Drs. Noriyuki Uchida of AIST and Yuji Ohishi of Osaka University for providing the opportunity to use the κ measurement system and helpful discussions. This work is partly supported by PRESTO (No. JPMJPR15R2) and CREST (No. JPMJCR19Q5) from the JST in Japan, JSPS KAKENHI (Nos. 19K21971, 19H00853, 20H05188, and 21H01366), and a research grant (Creation of Life Innovation Materials for Interdisciplinary and International Researcher Development) from the MEXT in Japan.

References

1. S. N. Khatami and Z. Aksamija, *Appl. Phys. Rev.* **6**, 014015 (2016).
2. M. Kurosawa, M. Fukuda, K. Takahashi, M. Sakashita, O. Nakatsuka, and S. Zaima, "Thermophysical characterizations of $Ge_{1-x}Sn_x$ epitaxial layers aiming for thermoelectric devices," in *Abstr. Book of 9th Int. Conf. on Silicon Epitaxy and Heterostructures*, 2015, p. 123.
3. D. Spirito, N. von den Driesch, C. L. Manganelli, M. H. Zoellner, A. A. C. -Wiciak, Z. Ikonic, T. Stoica, D. Grützmacher, D. Buca, and G. Capellini, *ACS Appl. Energy Mater.* (2021). (Article ASAP, DOI: 10.1021/acsaem.1c01576)
4. N. Uchida, J. Hattori, R. R. Lieten, Y. Ohishi, R. Takase, M. Ishimaru, K. Fukuda, T. Maeda, and J.-P. Locquet, *J. Appl. Phys.* **126**, 145105 (2019).
5. M. Kurosawa, K. Liu, M. Izawa, I. Tsunoda, S. Zaima, *ECS Trans.* **75**(8), 481 (2016).
6. K. Takahashi, H. Ikenoue, M. Sakashita, O. Nakatsuka, S. Zaima, and M. Kurosawa, *Appl. Phys. Express* **12**, 051016 (2019).
7. K. Sato, O. Nakatsuka, and M. Kurosawa, "Influence of Dopant on Thermoelectric Properties of Si-rich Poly-$Si_{1-x}Sn_x$ Layers Grown on Insulators," in *Ext. Abstr. of 2019 Int. Conf. on Solid State Devices and Materials*, 2019, p. 1009.
8. Y. Peng, L. Miao, J. Gao, C. Liu, M. Kurosawa, O. Nakatsuka, and S. Zaima, *Sci. Rep.* **9**, 14342 (2019).
9. Y. Peng, H. Lai, C. Liu, J. Gao, M. Kurosawa, O. Nakatsuka, T. Takeuchi, S. Zaima, S. Tanemura, and L. Miao, *Appl. Phys. Lett.* **117**, 053903 (2020).
10. T. Iwahashi, M. Kurosawa, N. Uchida, Y. Ohishi, T. Maeda, O. Nakatsuka, and S. Zaima, "Sb-doping effect on thermal and electrical properties of Ge-rich $Ge_{1-x}Sn_x$ layers," in *Ext. Abstr. of 2017 Int. Conf. on Solid State Devices and Materials*, 2017, p. 596.
11. Y. Imai, K. Takahashi, N. Uchida, T. Maeda, O. Nakatsuka, S. Zaima, and M. Kurosawa, "Domain size effects on thermoelectric properties of p-type $Ge_{0.95}Sn_{0.05}$ layers grown on GaAs and Si substrates," in *Proc. of 2018 IEEE 2nd Electron Devices Technology and Manufacturing Conference*, 2018, p. 310.
12. C. J. Glassbrenner and G. A. Slack, *Phys. Rev.* **134**, A1058 (1964).
13. S. Zaima, O. Nakatsuka, N. Taoka, M. Kurosawa, W. Takeuchi, and M. Sakashita, *Sci. Technol. Adv. Mater.* **16**, 043502 (2015).

14. M. Kurosawa, M. Kato, K. Takahashi, O. Nakatsuka, and S. Zaima, *Appl. Phys. Lett.* **111**, 192106 (2017).
15. J. Callaway and H. C. von Baeyer, *Phys. Rev.* **120**, 1149 (1960).
16. P. G. Klemens, *Proc. R. Soc. Lond. A* **208**, 108 (1951).
17. G. A. Slack and S. Galginaitis, *Phys. Rev.* **133**, A253 (1964).
18. A. J. Minnich, H. Lee, X. W. Wang, G. Joshi, M. S. Dresselhaus, Z. F. Ren, G. Chen, and D. Vashaee, *Phys. Rev. B* **80**, 155327 (2009).
19. K. Kato, Y. Hatasako, M. Kashiwagi, H. Hagino, C. Adachi, and K. Miyazaki, *J. Electron. Mater.* **43**, 1733 (2014).
20. T. Watanabe, S. Asada, T. Xu, S. Hashimoto, S. Ohba, Y. Himeda, H. Zhang, M. Tomita, and T. Matsukawa, *Proc. of 2018 IEEE 2nd Electron Devices Technology and Manufacturing Conference*, 2017, p. 86.
21. M. Tomita, S. Ohba, Y. Himeda, R. Yamato, K. Shima, T. Matsuki, and T. Watanabe, *IEEE Trans. Electron Devices* **65**, 5180 (2018).
22. A. Toriumi and T. Nishimura, *Jpn. J. Appl. Phys.* **57**, 010101 (2018).
23. K. Yamamoto, R. Noguchi, M. Mitsuhara, M. Nishida, T. Hara, D. Wang, and H. Nakashima, *J. Appl. Phys.* **118**, 115701 (2015).
24. M. Koike, Y. Kamimuta, and T. Tezuka, *Appl. Phys. Express* **4**, 021301 (2011).
25. J. Jeon, A. Suzuki, O. Nakatsuka, and S. Zaima, *Semicond. Sci. Technol.* **33**, 124001 (2018).
26. T. Maeda, W. H. Chang, T. Irisawa, H. Ishii, H. Oka, M. Kurosawa, Y. Imai, O. Nakatsuka, and N. Uchida, *Semicond. Sci. Technol.* **33**, 124002 (2018).
27. W. H. Chang, T.-Z. Hong, P.-J. Sung, T. Irisawa, H. Ishii, Y.-J. Lee, and T. Maeda, *ECS Trans.* **102**(4), 17 (2021).

190

Chapter 8

Novel Process-Nanofabrication

Study on Impact of MOS Interface Passivation Processes on Band Alignment and Flat-Band Voltage of 4H-SiC Gate Stacks

Koji Kita

Department of Materials Engineering, The University of Tokyo
7-3-1 Hongo, Bunkyo-ku, Tokyo 113-8656, Japan

Change of band alignment at SiO_2/4H-SiC was observed by extending the duration of interface nitridation process to passivate the interface defects. The conduction-band offset on (0001) stack increased but that on (000-1) and (1-100) stacks decreased, which did not result only in a difference of flatband voltage (V_{fb}) but also in a difference of gate leakage characteristics among those stacks on different crystal faces. An introduction of an additional dipole layer in the gate stack was demonstrated by depositing a few-nm-thick Al_2O_3 on SiO_2 to cause a positive shift of V_{fb} without degrading the quality of SiO_2/SiC MOS interface. Control of those factors to cause V_{fb} shift is crucially important to reduce the mobility degradation in SiC MOSFET inversion channel by increasing the channel doping concentration to tune the threshold voltage.

Introduction

4H-SiC is one of the widegap semiconductors suitable for power device applications and its mass production technologies has been rapidly developed. A sufficiently large threshold voltage (V_{th}) is generally demanded for high-voltage power MOSFETs to avoid the malfunction in the operations under large electrical noises. However, in the case of 4H-SiC MOSFETs, the increase in channel doping concentration to tune V_{th} higher inevitably induces a significant reduction of carrier mobility in the inversion channel, due to the dominant Coulomb scattering caused by defective SiO_2/4H-SiC interface [1]. Thus the technique to tune the threshold voltage is crucial for SiC MOSFETs. Another serious problem is that negative V_{th} shift is often observed by employing SiC surface nitridation process to passivate the interface defects, usually conducted by a post-oxidation annealing (POA) in NO ambient at elevated temperature [2]. Such processes have been found to introduce nitrogen atoms selectively to the surface of SiC, by substituting the topmost carbon atoms. The surface of SiC is stabilized typically by the formation of nitrogen atoms coordinated by three Si atoms [3, 4]. Such nitrogen passivation process is inevitable to fabricate 4H-SiC MOS gate stacks with sufficiently low interface state density and good device reliability. Therefore, the development of the technique to design V_{th} of 4H-SiC MOSFETs by overcoming unwanted negative shift caused by NO-POA is strongly demanded. In this study we investigate the factors to determine the flatband voltage (V_{fb}) of 4H-SiC gate stacks to understand the guideline to reduce the channel mobility degradation caused by the increase of channel doping concentration to tune the V_{th} in 4H-SiC devices.

Experiments

After chemical cleaning with diluted HF solution, n-type 4H-SiC substrates, covered with ~5 μm-thick epitaxial layer (~1×10^{16} cm^{-3} n-type doped) on different crystal orientations: (0001) Si-face, (000-1) C-face, and (1-100) m-face, were thermally oxidized at 1300°C for various durations to obtain thermal oxides with different thicknesses, followed by POA in NO + N_2 ambient at 1150°C for different durations from 0 to 8 hrs. For some of the samples ~3 nm-thick or ~6 nm-thick Al_2O_3 layer was additionally deposited by rf-sputtering followed by annealing at 800°C in 0.1% O_2 + N_2 ambient. Finally, Au electrode was deposited to fabricate MOS capacitors. To determine the band alignment two methods were employed: the estimation from Fowler-Nordheim (F-N) tunneling current and that from valence-band spectrum analysis in x-ray photoelectron spectroscopy (XPS).

Impact of NO Annealing on Band Alignment

Change of band alignment by interface nitridation

First we systematically studied the V_{fb} of SiO_2/4H-SiC (0001) n-type MOS capacitors with various oxide thickness for different NO-POA durations as shown in Fig. 1. From the oxide thickness dependence of V_{fb} we could deduce the "expected V_{fb}" value when the effects of fixed charges are removed, from the offset of Fig. 1 by extrapolating the thickness-dependence of V_{fb} to zero-thickness. Note that if V_{fb} shift is caused only by fixed charge introduction, the offset value should be consistent with the ideal V_{fb} determined simply by the Fermi level difference between the gate electrode and the semiconductor. However, what we found from the series of samples was an anomalous negative shift of the offset values by changing the NO-POA duration in several hundreds of mV, which was enlarged by extending NO annealing duration [5] as shown in the figure. This phenomenon is quite unexpected but would be explainable by assuming a growth of dipole layer at the interface by SiC surface nitridation. Considering the polar Si–N bond formation aligned on single crystal surface, the passivated SiC surface may induce an additional potential change. From the intensity ratio of N1s to Si2p core-level XPS observed after removal of SiO_2 layer, high density of nitrogen ~10^{14} cm^{-2} introduction at the surface of SiC after 8hr NO-POA was estimated on (0001) surface. As mentioned above, the introduced nitrogen atoms are expected to occupy mainly the topmost C-sites [3, 4] to be coordinated by three Si-atoms on SiC side (N–Si$_3$). Then we can expect a generation of dipole moment normal to the surface with such an asymmetric structure. Considering the alignment of array of N-Si$_3$ structures at the interface with high density, it would not be surprising to observe a significant dipole effect by the formation of nitrided surface on SiC.

Note that contributions of the fixed charges to the V_{FB} shifts in our samples on (0001) Si-face substrates were quite small after sufficiently long NO-POA, as can be seen in the very small slope in Fig. 1. On the other hand, we found an introduction of interface fixed charges only when we conducted an additional annealing at high temperature >1000°C after NO-POA (data not shown), probably attributed to a partial decomposition of nitrogen-terminated structure at the interface to produce trap states. In general, SiC MOSFET fabrication processes often employ a high temperature post-metallization annealing processes which might cause a serious increase of fixed charge density. From these results we can conclude that the unwanted negative shift of V_{fb} often reported for SiC MOS devices with nitridation processes is consisting of two components: the dipole layer formation by introduction of nitrogen and the fixed charge formation mainly by the partial decomposition of nitrided surface structure by additional annealing process after NO-POA.

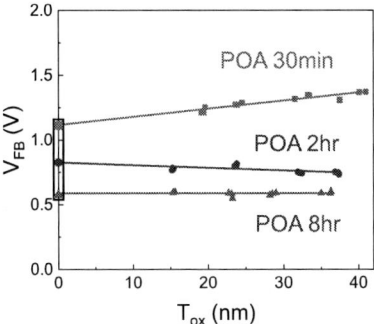

Figure 1. SiO_2 thickness dependence of V_{fb} of SiO_2/4H-SiC (0001) MOS capacitors fabricated with various NO-POA conditions.

Next, the SiO_2/4H-SiC band alignment was investigated by two kinds of methods: the valence-band spectra analysis using XPS and the energy barrier height characterization of F-N current analysis [5]. For the former analysis we obtained valence-band spectra of a sufficiently-thick-SiO_2 film and a SiC substrate without SiO_2 formation, together with the SiO_2/SiC stacks with ~3 nm-thick SiO_2 where the valence bands of both SiO_2 and SiC were observable simultaneously. Then by deconvoluting the spectrum of ~3 nm-thick SiO_2 sample into the components of SiO_2 and SiC, to evaluate the offset of the valence band edge. Then assuming the bandgap of SiO_2 as 8.7 eV and that of SiC as 3.26 eV, the conduction band offset was estimated [5]. For the F-N current analysis, we characterized leakage current at −150°C where other current conduction mechanisms like Pool-Frenkel (PF) current was suppressed. The energy barrier height appeared in the equation of F-N tunneling is regarded as the conduction-band offset at SiO_2/SiC. From these analyses, as shown in Fig. 2 (a), a clear trend was observed that the valence-band offset between 4H-SiC (0001) and SiO_2 changes monotonically by extending NO-POA durations, to result in an increase of conduction-band offset at SiO_2/SiC interface, which naturally explains the unexpected negative shift of V_{fb} observed in MOS capacitors as discussed above.

When we compared the band alignment change by NO-POA for the stacks on the substrates with different crystal orientations, to our surprise, these trends appeared to the opposite direction for the gate stacks on 4H-SiC (000-1) to that we observed on (0001) as shown in Fig. 2 (b), even though they were all fabricated with similar oxidation conditions and identical nitridation processes. For (1-100) m-face stacks, the trend was quite similar to that of (000-1) (data not shown). By extending the NO-POA durations, (0001) stack shows the increase of conduction-band offset (reduction of valence-band offset), whereas (000-1) and (1-100) stacks show the decrease of conduction-band offset. This contrast is indicating the difference among the dipole directions formed at those nitrided interfaces. For (000-1) stacks, the opposite direction of the dipole moment to that on (0001) would be explainable by the formation of N–Si$_3$ structure upside down [5]. For (1-100) m-face stacks, however, the physical origin of the dipole moment is not so easy to be explained with such a simple model, because the interface structure of SiO$_2$/4H-SiC (1-100) [6] has not been understood well and the dominant sites of nitrogen at the interface is not clearly determined yet.

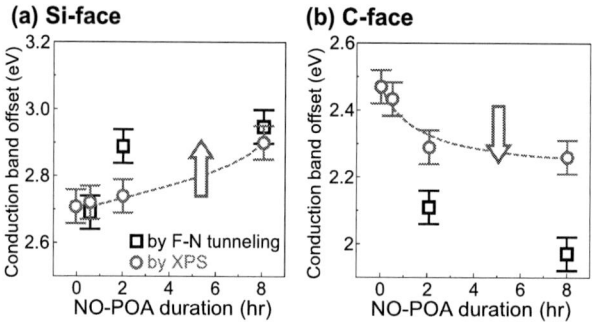

Figure 2.   Change of estimated conduction-band offset of SiO$_2$/4H-SiC stacks by extending NO-POA annealing duration for the stacks on (a) (0001) Si-face and (b) (000-1) C-face.

Change of leakage current levels by interface nitridation

The gate leakage current level is a critical factor to determine the reliability of the gate stacks. Especially, understanding of the leakage characteristics on (1-100) m-face would be important from the viewpoint of device applications, considering the recent development of trench-type MOSFET technologies where the vertical sides of the trench on (0001) wafers are utilized for the channel. As discussed above, the direction of the band alignment shift caused by the nitridation process on (1-100) is beneficial to achieve a higher V$_{fb}$, however, a serious drawback would be the reduction of conduction band offset. Actually the F-N tunneling current levels observed at −150°C, which dominantly determined by the conduction band offset were significantly different for the stacks on different crystal orientations. The F-N tunneling current is gradually decreasing on (0001) Si-face by extending NO-POA whereas increasing on (1-100) m-face, which is consistent with the current model of band alignment change. At room temperatures or higher, on the other hand, the defect-assisted conduction such as P-F model emerges in addition to the F-N tunneling current. In this study we assumed the observed leakage current at high temperature is given by the sum of F-N tunneling and P-F emission current. Then the P-F current can be separately characterized by the subtraction of F-N current from the observed total leakage current at each temperature, by assuming a certain temperature dependence of F-N current [7]. As a result, we could confirm that the extracted component well followed the P-F model by assuming the relative optical dielectric constant ~2 for SiO$_2$ and the trap level of 1.4~1.6 eV below the conduction band edge, which is consistent with previous report of P-F current in SiC gate stacks [7].

In Fig. 3 the P-F current and F-N current components observed at 100°C are shown as a function of oxide electric field, for the stacks on (0001) and (1-100) fabricated with

NO-POA for 8hr. As can be seen in the figure, the P-F current dominates especially for higher electric field region at this temperature. It should be noted that P-F component, as well as F-N current, was much higher for (1-100) stack than (0001) stack at a given electric field. Even though the energy level of traps extracted from the P-F current analysis seems common for those stacks, the current level is considered to be affected significantly by the difference of the conduction-band offset between them. From these results, the band alignment change induced by interface nitridation process works disadvantageous for (1-100) m-face in terms of leakage current level. Even though there is a clear advantage of employing trench channel configuration than planar channel to reduce the resistance of the SiC power MOSFETs, the effects of band alignment difference between (1-100) and (0001) stacks induced by NO-POA should be taken into consideration to discuss the difference of gate stack characteristics between trench and planar channel devices.

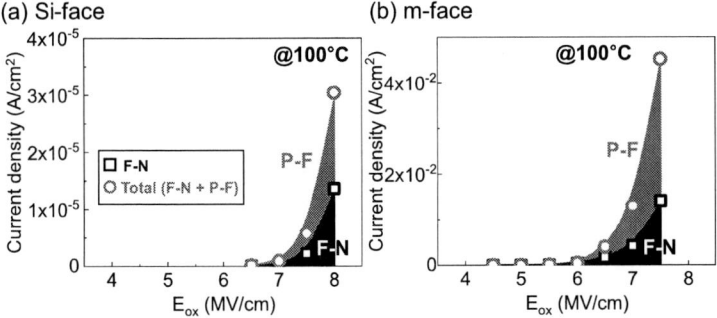

Figure 3. Estimation of F-N tunneling and P-F emission current components from the leakage current measured at 100°C shown as a function of electric field in oxide for SiO_2/4H-SiC MOS capacitors on (a) (0001) Si-face and (b) (1-100) m-face substrates.

Introduction of additional dipole layer to manipulate flatband voltage

As we have discussed so far, V_{fb} of SiC gate stacks are affected by dipole layer formation at SiO_2/SiC interface, in addition to the positive fixed charge generation at the interface mainly during the additional annealing process after interface nitridation. To achieve a more positive shift of V_{fb}, in this section we examine a way to intentionally introduce another dipole layer in the stack, by depositing a few nm-thick Al_2O_3 layer on top of SiO_2 based on the concept of oxide interface dipole layers, which has been reported for various high-k/SiO_2 interfaces [8].

We deposited ~3 nm-thick or ~6 nm-thick Al_2O_3 layers by rf-sputtering on the SiO_2/4H-SiC (0001) stacks after NO-POA for 2hr, followed by a post-deposition annealing (PDA) in 0.1% $O_2 + N_2$ ambient at 800°C. This PDA temperature was chosen taking account of the finding that a higher temperature annealing will induce a positive fixed charge at the interface, as we have already discussed. As a result, we successfully demonstrated 0.6~0.8V positive shift of V_{fb} by an introduction of thin Al_2O_3 layers [9], as shown in Fig. 4. Since we observed a similar V_{fb} for both ~3nm and ~6nm stacks, such positive shift of V_{fb} should not be attributed to a fixed charge introduction but to the

Al_2O_3/SiO_2 interface dipole layer formation. It should be noted that those processes of Al_2O_3 deposition and PDA at 800°C do not affect the interface quality, in terms of interface defect density (D_{it}) characterized by the conductance method (data not shown) [9]. In this study we only demonstrated a single layer of Al_2O_3 deposition, but a possible approach to achieve a further positive V_{fb} shift will be using a laminated stack with a few-nm-thick Al_2O_3 and SiO_2 repetition to enlarge the strength of dipole layer, as was demonstrated on Si substrate stacks [10]. This technique is expected as a method not only to overcome the unexpected change of V_{fb} caused by the nitrogen passivation process or the additional annealing after nitridation, but to achieve higher V_{th} without increasing the channel doping concentration. Since a serious deterioration of channel mobility is expected for higher channel doping concentration in SiC MOSFETs due to the dominance of Coulomb scattering by the defect states locating at the interface, such technique will help to design a channel to achieve higher mobility with sufficiently high V_{th}.

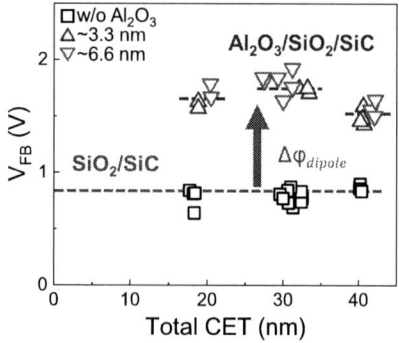

Figure 4. V_{fb} of $SiO_2/4H$-SiC and $Al_2O_3/SiO_2/4H$-SiC (0001) MOS capacitors as a function of capacitance equivalent thickness, which clearly indicates the shift of V_{fb} induced by an additional dipole layer formation at Al_2O_3/SiO_2 interface.

Conclusions

We found that nitrogen passivation of $SiO_2/4H$-SiC MOS interface defects resulted in a significant change of band alignment to increase the conduction band offset on (0001) stack but to decrease the offset on (000-1) or (1-100) stacks, which resulted in a difference of gate leakage characteristics as well as V_{fb}. An additional dipole layer can be introduced in the gate stack by depositing thin layer of different dielectric material on top of SiO_2. The formation of dipole layer by depositing Al_2O_3 on SiO_2 resulted in a positive shift of V_{fb} without degrading the quality of SiO_2/SiC MOS interface. The effects of dipole layers formed at SiO_2/SiC interface and/or in the multilayer dielectrics would have significant roles to tune V_{th} of SiC MOSFETs. Such guidelines to tune V_{fb} without changing the channel doping concentration would contribute to improve SiC MOSFET inversion channel mobility which is often severely degraded by increasing the channel doping concentration to adjust V_{th}.

Acknowledgments

This work was done in collaboration with Tae-Hyeon Kil in the University of Tokyo and partly supported by JSPS KAKENHI Grant Numbers 18H03771 and 21H04550. The author is grateful to Dr. Munetaka Noguchi and Dr. Hiroshi Watanabe in Mitsubishi Electronics for fruitful discussions, and Dr. Masat Noborio and Dr. Sumera Shimizu in DENSO Corporation for their kind advices for our study.

References

1. M. Noguchi, T. Iwamatsu, H. Amishiro, H. Watanabe, K. Kita, and S. Yamakawa, *Jpn. J. Appl. Phys.* **58**, 031004 (2019).
2. J. Rozen, S. Dhar, M. E. Zvanut, J. R. Williams, and L. C. Feldman, *J. Appl. Phys.* **105**, 124506 (2009).
3. T. Shirasawa, K. Hayashi, S. Mizuno, S. Tanaka, K. Nakatsuji, F. Komori, and H. Tochihara, *Phys. Rev. Lett.*, **98**, 136105 (2007).
4. Y. Xu, X. Zhu, H. Lee, C. Xu, S. Shubeita, A. Ahyi, Y. Sharma, J. Williams, W. Lu, and S. Ceesay, *J. Appl. Phys.* **115**, 033502 (2014).
5. T. H. Kil and K. Kita, *Appl. Phys. Lett.* **116**, 122103 (2020).
6. D. Mori, Y. Fujita, T. Hirose, K. Murata, H. Tsuchida, and F. Matsui, *Appl. Phys. Lett.* **112**, 131603 (2018).
7. M. Sometani, D. Okamoto, S. Harada, H. Ishimori, S. Takasu, T. Hatakeyama, M. Takei, Y. Yonezawa, K. Fukuda, and H. Okumura, J. Appl. Phys. 117, 024505 (2015).
8. K. Kita and A. Toriumi, *Appl. Phys. Lett.* **94**, 132902 (2009).
9. T. H. Kil, M. Noguchi, H. Watanabe, and K. Kita, *Solid-State Electronics* **183**, 108115 (2021).
10. H. Kamata and K. Kita, *Appl. Phys. Lett.* **110**, 102106 (2017).

Modeling of Advanced FinFET Dummy Gate Corner Residue Impacted by Clogging

X. Xiao[a], X. Ke[a], Bo Su[a], and H. Zhang[a]

[a] Semiconductor Manufacturing International Corp., Shanghai 201203, China

> In this work, we model 3D corner residues based on a process flow simulation using the Coventor SEMulator3D virtual platform. The role of clogging existing during plasma etch process in 3D corner formation has been studied based on the simulation results. In particular, the impacts of clogging and plasma aspect ratio distribution functions on corner size in typical etching steps are assessed. Furthermore, tunable model provides insights on the effect of fin height variation. Higher Fin structure could provide more shadowing effects which encourage the aggregation of 3D corner residue. In an advanced model containing multiple dry etch steps, the window of plasma divergence is also discussed, which is helpful to study and deal with the existing of 3D corner residue.

Introduction

Owing to the distinct properties such as fast switching speed, lower power consumption and better device properties, FinFET technology has been applied in advanced nodes below 22 nm instead of planar MOSFET (1-3). In a typical FinFET manufacturing flow, despite the following isolation and metallization process, device performance will be impacted by dummy gate definition, including patterning and dry etch process. However, poly etch processes are challenged by the existing of 3-Dimensional (3D) corner residue, which is widely found at the intersection between Gate and Fin in advanced FinFET technologies (4-6). Commonly, the 3D corner residue as a key geometry for device yield and reliability is induced by plasma shadowing effect caused by Gate and Fin structure during dummy gate etch process. Clogging on the top of Fin structure formed during complex gate etch process is believed to play an important role on the shadowing effects inducing 3D corner residue.

The possible sources of clogging include deposition behavior of polymer formed during gate etch process, and subsequent sputtering of plasma containing ions and radicals from backside. It is obvious that more 3D corner residue would be induced by larger clogging size, while the certain relationship between the two aspects remain unclear due to the difficulty of measurement in situ during etch process. Simulation methods could be the ideal candidate to monitor the possible forming mechanism of 3D corner residue with the shadowing effect of clogging in the absence of measurements in situ.

In this paper, we proposed a method to measure and study the 3D corner residue by modeling the dummy gate etch process of advanced FinFET technology via Coventor SEMulator3D virtual platform. The main goal is to take an insight on the correlations of clogging and 3D corner, together with the possible impact of fin height variation. The process flow simulation and characterization methodology will be presented subsequently, and the model is applied to simulate Fin height variability and characterize the consequences of 3D corner residue.

Metrology

A conventional dummy gate etch flow is simulated utilizing Coventor SEMulator3D and described in Figure 1. The Fin structure is formed firstly. After dep of poly and hard mask layers, the gate mandrel is patterned. A nitride spacer is deposited uniformly and etched, before the core is chemically removed. The remaining spacer acts as a hard mask for the dry etching step of the underlying amorphous silicon layer. Then the pattern is transferred into the silicon substrate, leaving the silicon oxide and nitride hard mask on top of the dummy gate. The dummy gate etch step mainly consists of main etch, soft landing and over etch steps. During soft landing, relatively lower RF bias power with abundant ions or radicals leads to the formation of clogging on top of Fin, which functions as an "umbrella" to prevent the plasma to reach the bottom of Fin and gate in the following over etch step. In this work, the plasma behavior during over etch step is mainly studied in the presence of clogging with different sizes.

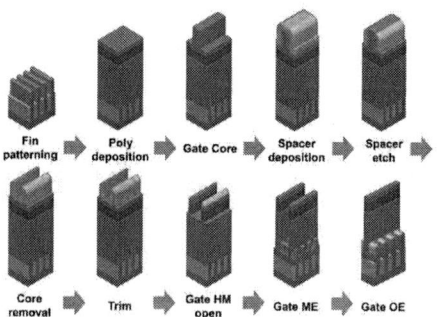

Figure 1. Process flow of dummy gate patterning.

The model is simplified and only two factors, clogging and aspect ratio distribution functions (ARDF) have been considered as the tuning parameters with other variables fixed. To simulate the clogging structure with different sizes, deposition by cycle is introduced. In detail, the clogging size split is realized by adding cycles with a small amount per cycle. The clogging profile is shown in Figure 2. Little steric hindrance could be observed at first, while the Fin space decreases sharply with adding cycles. Thus, the plasma containing ions and radicals can hardly reach the bottom of Fin and gate. ARDF with a series of values is applied in this model to obtain the final 3D corner residue profile. Considering the high RF bias power applied in production, the range of ARDF is set from 0 to 0.1 in the zone of highly directional source.

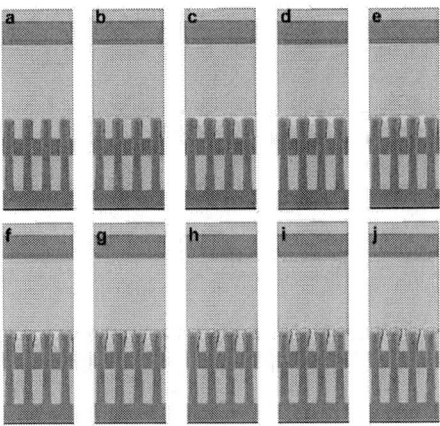

Figure 2. Clogging formation schematic illustration with adding cycles from original structure (a) to (j).

The main parameters to be considered are the size of clogging on Fin top and the formed 3D corner residue at the bottom of Fin/gate. The measurement is conducted on the plane right between two adjacent gate. The Fin pitch, Fin top and bottom CD are considered as constant and the measurement takes place both at Fin top and bottom sites. The size of clogging and 3D corner residue are calculated as below:

$$Clogging = (Fin \ pitch-Fin \ top \ CD-CD1)/2 \qquad [1]$$
$$Residue = (Fin \ pitch-Fin \ Bottom \ CD-CD2)/2 \qquad [2]$$

This enables a reliable measurement of clogging and 3D corner residue for any process variation simulated.

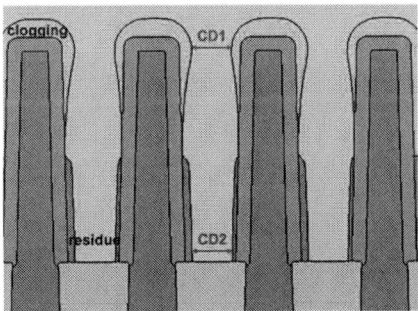

Figure 3. Clogging size and 3D corner residue measurement: measure the CD of dummy material filled in the space at the Fin top and bottom sites. The clogging size and residue could be calculated by the value measured and Fin information.

Results and Discussions

The model described earlier is applied to study the performance of 3D corner residue with a fixed FinFET structure. The plasma ARDF varies from 0 to 0.1 with the existing of different clogging size (0.5 to 4.21 as calculated). The 3D corner size obtained is depicted in Figure 4 as a function of the two parameter dimensions mentioned above. It is obvious that smaller corner size could be obtained with less clogging and higher directional source as predicted. For example, with a relatively small corner size (<1.5 nm) and narrow ion angle distribution (<0.04), the corner sites can almost be reached by plasma with only little residue remaining. However, we noticed that the ARDF plays a more significant role in the

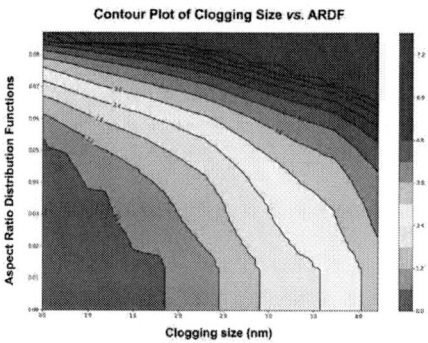

Figure 4. Contour plot of clogging size versus aspect ratio distribution functions.

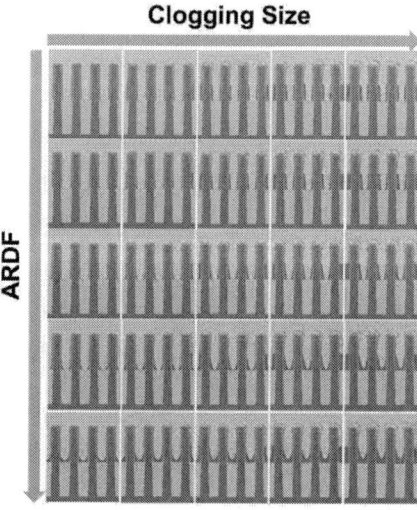

Figure 5. Cross-section view of 3D corner residue structures with increasing clogging size and plasma ARDF.

formation of 3D corner for the corner size rises sharply as the ARDF ranges from 0.06 to 0.08. Meanwhile, no difference in corner size could be observed with ARDF lower than 0.01, indicating the shadowing effect induced by clogging on Fin top. In order to obtain a direct insight of the clogging and corner residue profile, the cross-section images of 25 splits with multiple conditions are summarized in Figure 5. With relatively large ARDF, even the bottom sites far from dummy gate footing remain undamaged by plasma, indicating the sufficient shadowing effects of clogging. It is worthy to notice that 3D corner residue remains in the condition of no clogging and large ARDF value, indicating the significant effect of the aspect ratio of Fin height versus dummy gate pitch. In the case that dummy gate pitch remains constant, the corner size is suspected to vary with Fin height.

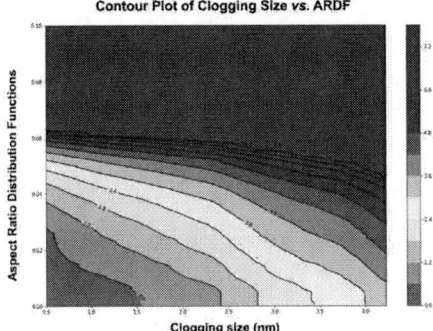

Figure 6. Contour plot of clogging size versus aspect ratio distribution functions with larger Fin height.

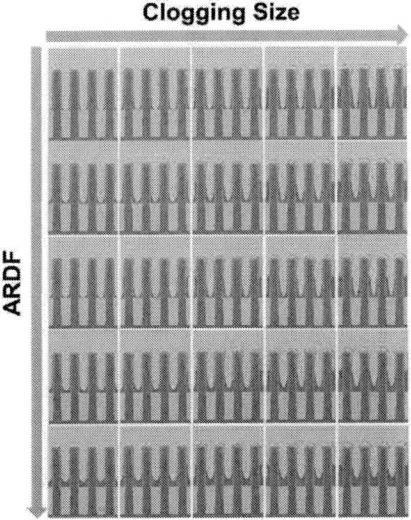

Figure 7. Cross-section view of 3D corner residue structures with larger Fin height.

Nevertheless, Fin height is also taken into account to study the impacts of aspect ratio of depth and critical dimension (CD). As shown in Figure 6, with higher Fin structure, the space between adjacent dummy gate is almost filled with unreacted components in the case of ARDF higher than 0.06, regardless the value of clogging sizes. This results reveal that the aspect ratio of the 3D space surrendered by Fin and dummy gate can define the difficulty for ions and radicals to reach the bottom. Similar contour plot could be observed with lower ARDF and the corner size is not sensitive with the increase of clogging size, possibly due to the higher aspect ratio of the Fin and dummy gate structures. As shown in the cross-section images (Figure 7), the bottom cannot be reached by plasma with higher ARDF, resulting in plenty of residue.

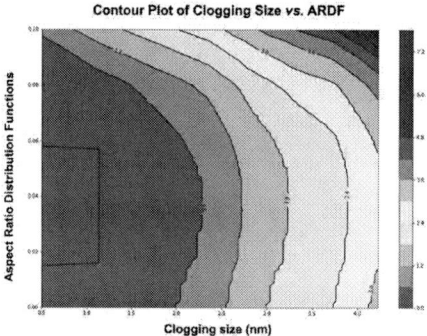

Figure 8. Contour plot of clogging size versus aspect ratio distribution functions with smaller Fin height.

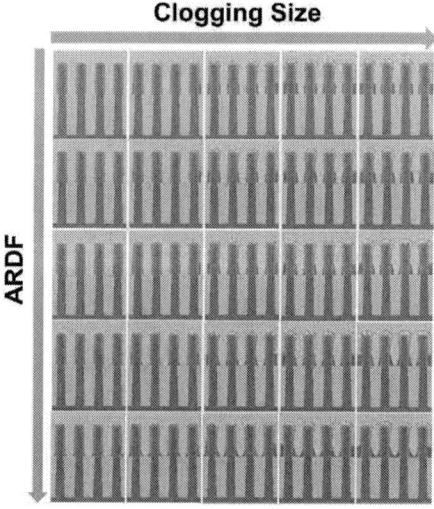

Figure 9. Cross-section view of 3D corner residue structures with smaller Fin height.

In the opposite situation, better 3D corner performance could be obtained with lower Fin structure (Figure 8). Corner residue can almost be eliminated even with higher ARDF or larger clogging due to the low aspect ratio of Fin-dummy gate structure, which weaken the shadowing effects of clogging formed. It is worthy to notice that in the case of uniform clogging size, smaller corner residue can be observed with larger ARDF within a certain range of 0-0.04. A possible explanation is the shadowing effect is too weak so that only plasma with a little incident angle could reach the dummy gate footing to fully eliminate residues. Similar situation could be observed in Figure 9, the formed corner residues are mainly behind clogging instead of the middle of dummy gate space.

Conclusion

In this work, a FinFET structure involving 3D corner residues has been constructed using the Coventor SEMulator3D virtual platform. Based on the simulation, the impacts of clogging on top of Fin have been studied thoroughly, together with the dimension of plasma direction functions. Fin height is also taken into considerations to find out the potential effects. The constructed model containing multiple dry etch steps is helpful to study and deal with the existing of 3D corner residue, which provides insights to study the in-situ behavior of dummy gate patterning.

References

1. L. Liebmann, J. Zeng, X. Zhu, L. Yuan, G. Bouche and J. Kye, In *IEEE Symposium on VLSI Technology.* (2016).
2. A. P. Jacob, R. Xie, M. Sung, L. Liebmann, R. Lee and B. Taylor, *International Journal of High Speed Electronics and Systems*, **26**(01n02), 1740001(2017).
3. T. Dillinger. *Highlights of the TSMC Technology Symposium*, Part 1(2020).
4. Q. Han, X. Meng and H. Zhang, In *China Semiconductor Technology International Conference* (2015).
5. X. Zhang, M. Karakoy, K. Wu, Z. Chen, Z. Ge, N. Krishnan, A. Siany, S. Levi, I. Schwarzband and R. Kris, In *2018 29th Annual SEMI Advanced Semiconductor Manufacturing Conference (ASMC)*, 320(2018).
6. D. Dunn, J. R. Sporre, V. Deshpande, M. Oulmane, R. Gull, P. Ventzek and A. Ranjan, In *Advanced Etch Technology for Nanopatterning VI, International Society for Optics and Photonics*, 10149, 101490Q(2017).

Process Development of Dislocation-Free SiGe P-Channel in FinFET Technology

B. Su[a], W. F. Deng[a], C. Yin[a], E. N. Zhang[a], X. Ke[a], H. Oh[a], J. Zhao[a], B. Ye[a], and H. Y. Zhang[a]

[a] Semiconductor Manufacturing International Corporation, Shanghai 201203, China

SiGe channel metal oxide semiconductor field effect transistor (MOSFET) became a choice of material as a performance booster in leading edge logic technologies. In this study, we demonstrate integration solutions for dual Fin formation (Si for nFET, SiGe for pFET) using buried SiGe channel approach on 300 mm Si wafer. Plasma-induced damage (PID) layer, which thickness ranges from 0.5 nm to 2.5 nm, was observed after Si trench (p-trench) etch for SiGe EPI growth. This PID inorganic layer leads to SiGe dislocation defect. And careful sequencing and optimization of oxidation and wet removal steps were able to realize defect-free SiGe EPI channel along with a thin buffer SiGe layer. Besides, margin assessment of p-trench based buried SiGe approach was also presented in this study. Moreover, dual Fin etch was demonstrated via inductively coupled plasma (ICP) using ALD-like function. The depth loading between SiGe and Si Fin was tunable via varying repeated cycles of advanced pulsing step, and less than 40 A of depth loading can be achieved. The progress reported represents a major leap for SiGe channel integration and paves the way for massive production.

Introduction

SiGe/Si dual Fin is considered as next candidate for metal-oxide semiconductor field-effect transistor (MOSFET) in advanced node, which fulfills scaling-down requirement. Except of Fin pitch and contact-to-poly pitch scaling-down, device performance can be enhanced by introduction of Ge into PMOS Si [1-5], which shows significant increase of hole mobility, easier Vt engineering and better negative-bias temperature instability (NBTI) than Si. However, SiGe suffers more oxidized consumption during integration processes, Ge self-diffusion, N-P boundary critical dimension (CD) and profile loading via dual Fin plasma etch, Fin CD dependent strain relaxation variation, and threading dislocation defect for SiGe [6-9]. Replacing Si with SiGe for PMOS is still challenge from MOSFET integration perspective. Therefore, it is imperative to rationally design a novel integration scheme for SiGe channel to overcome trade-off as mentioned. In this work, we demonstrate a promising buried SiGe integration scheme on 300 mm Si wafer for massive production purpose. Buried SiGe integration scheme involves several key steps: (1) PMOS area trench etch (p-trench etch) to define later SiGe channel; (2) trench surface restoration, in which plasma induced damage layer is oxidized and removed by wet and remote plasma clean; (3) SiGe epitaxy growth with graded Ge fraction; (4) SiGe planarization followed by Si capping layer and SiO2 deposition; (5) self-aligned patterning process; (6) dual Fin etch and SiN protection layer conformal deposition; (7)

STI oxide CMP and Fin reveal. To achieve SiGe and Si dual Fin etch with less depth loading, it is still a challenge due to relative higher etch rate for SiGe than Si using halogen chemistries plasma. Retaining of SiGe with around 0.3 concentration of Ge, this scheme substantially decreased dislocation defect in SiGe film, as well as excellent Si trench surface restoration and integration friendly with previous production tools, used for only Si channel device, regarding Ge introduced contamination. Furthermore, less than 40 A of depth loading between Si and SiGe Fin. Fully strained SiGe Fin as theoretical value is confirmed by nanobeam diffraction (NBD).

Process Integration

Key steps of buried SiGe integration scheme are shown in Figure 1. Controlling the surface damage and amorphous layer during plasma etch process is critical for p-trench etch. Therefore, special designed etch recipe, which is ended up with surface reaction layer removal steps, is applied for p-trench etch.

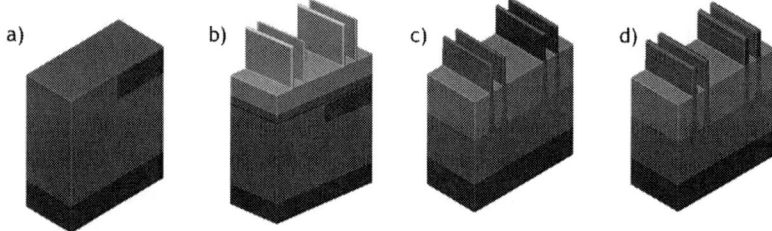

Figure 1. Key steps of buried SiGe integration flow with illustrations. a) post EPI Si cap; b) post SAQP patterning; c) dual Fin etch; d) Fin Reveal

Nevertheless, the plasma induced damage (PID) layer after p-trench etch is still observed via TEM, as shown in Figure 2. The plasma damaged layer contains sulfur instead of etch byproduct polymer. Furthermore, relative thicker damaged layer presents in trench bottom corner area. It indicates insufficient clean for PID layer removal via dedicated dry etch protocol and sequentially conventional dHF wet clean.

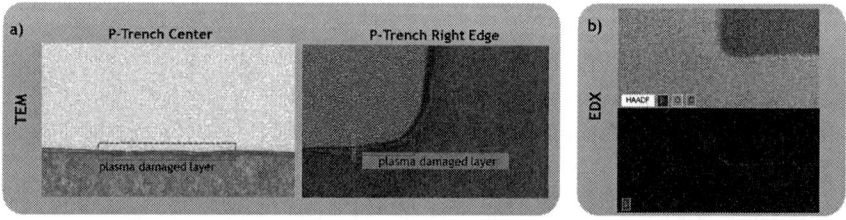

Figure 2. a) High resolution TEM of p-trench post conventional wet clean: trench center (left) and trench edge (right). Red dash rectangular area shows plasma damaged layer on Si surface. b) EDX results on trench bottom corner area. Sulfur element is detected on trench surface.

In case of without additional surface restoration steps, although prolong process time of conventional wet clean and pre-EPI clean treatment, SiGe crystal dislocation still occurs on interface of Si and SiGe, as well as SiGe bulk (Figure 3). Consequently, the additional step to restore PID layer is required. In this paper, in-situ steam generation (ISSG) oxidation process with low process temperature is applied to eliminate PID layer. Oxidation layer is further removed by diluted HF based wet chemical and remote plasma process, respectively. In this process, oxide wet etch amount needs to be well controlled, in order to avoid top Si exposure, which was initially covered by hard mask.

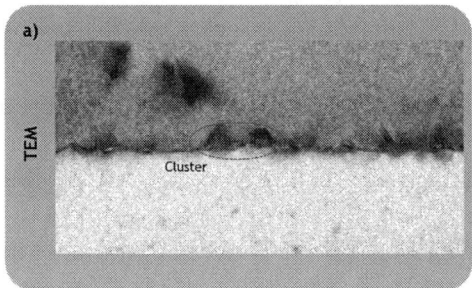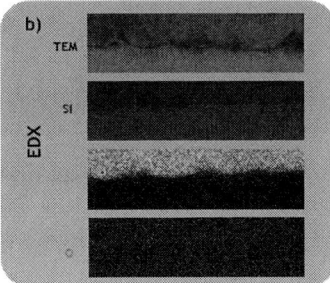

Figure 3. a) TEM of interface between epitaxy growth SiGe on P-Trench Si regime. Red dash circle area shows dislocation defect at Si and SiGe interface. b) EDX images, in which blue color represents Si element, yellow color represents Ge element, and purple represents O element.

In addition to PID layer removal, bottom corner radius (BCR) process window assessment based on ground rule could be calculated by formula below:

$$f(d, r, \theta) = (d - r) \cdot tan\theta + r \qquad [1]$$

$$\sigma^2(f) = tan^2\theta \cdot \sigma^2(d) + (1 - tan\theta)^2 \cdot \sigma^2(r) + [(d - r) \cdot \frac{1}{cos^2\theta}]^2 \cdot \sigma^2(\theta) \qquad [2]$$

where r represents BCR, σ is sigma value for each parameter, d is trench depth, a is distance between trench top to active area, θ is trench sidewall angle as shown in Figure 4.

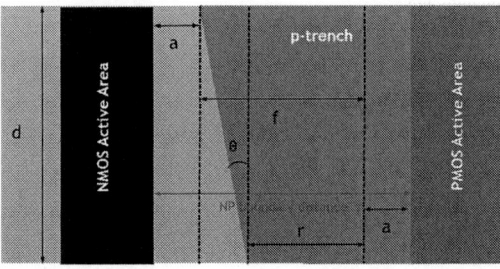

Figure 4. Illustration of p-trench process window assessment based on ground rule.

In order to deposit a high crystal quality of SiGe film, SiGe strain relaxed buffer (SRB) approach is successfully applied. Dislocation defects may be minimized via H_2 ambient anneal in each post-EPI growth. In Figure 5, high-resolution x-ray diffraction (HRXRD) data shows nearly consistent peak position with narrow full width at half maximum (FWHM) between wafer center and wafer edge, which confirms relaxation-free SiGe is obtained in wafer center and edge, respectively. Figure 6 ascribes p-trench area TEM images post SiGe CMP, in which SRB boarder is marked by read dots. It is noteworthy that BCR may enlarge in case of SRB application. In addition, N-P boundary process window is reduced as a result of SiGe SRB thickness. Therefore, it is important to control the buffer layer thickness on trench sidewall. It can be balanced by SiGe epitaxy growth rate along trench bottom (100) and sidewall (110) or using alternative sidewall protection films. Afterward, wet clean and EPI Si cap are followed to prevent SiGe oxidation during further patterning process. SiGe strain relaxation and Ge diffusion are suppressed by lower thermal budget in further integration flow.

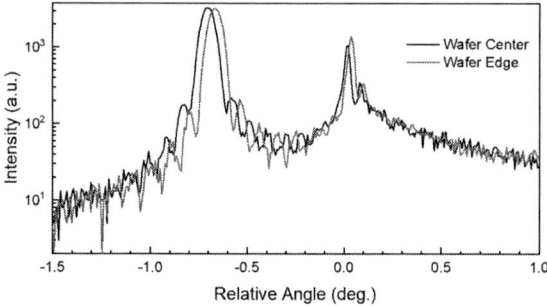

Figure 5. High resolution XRD spectra. Relaxation-free of SiGe is obtained at wafer center and edge, respectively.

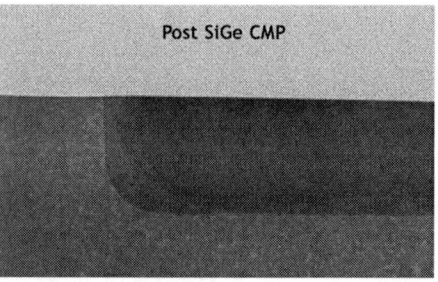

Figure 6. Cross-section of TEM image of P-Trench area post SiGe CMP. SiGe strain relaxed buffer boarder is marked by red dots.

To achieve SiGe and Si dual Fin etch with less depth loading, it is still a challenge due to relative higher etch rate for SiGe than Si using halogen chemistries plasma. Three reasons could be explained: (1) lower bonding energy of Si-Ge (3.12 eV) comparing with Si-Si (3.25 eV); (2) prior to remove strained material; (3) enhanced chemical reaction induced by narrower bandgap. Here, dual Fin plasma dry etch is performed by inductively coupled plasma (ICP) using ALD-like function to achieve comparable etch rate between Si and SiGe Fin, which is emphasized by Z-contrast STEM images as shown in Figure 7a. Fully strained SiGe Fin, along channel and perpendicular to wafer direction, is confirmed by NBD in Figure 7b. Controlling the depth loading between Si Fin and SiGe is crucial for device performance. Figure 8 indicates that depth loading, which not only refers to between Si and SiGe Fin, but also to pattern density, can be reduced by increasing repeated cycles of PET step, which include oxidation and halogen etch using advanced pulsing mode.

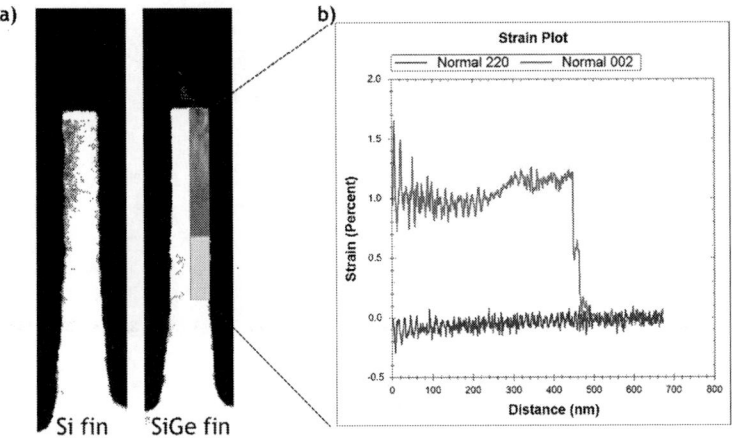

Figure 7. a) cross-section of Z-contrast STEM of Si (left) and SiGe (right) Fin post dual Fin etch step; b) Strain distribution of SiGe Fin along <220> channel direction and <002> perpendicular to wafer surface.

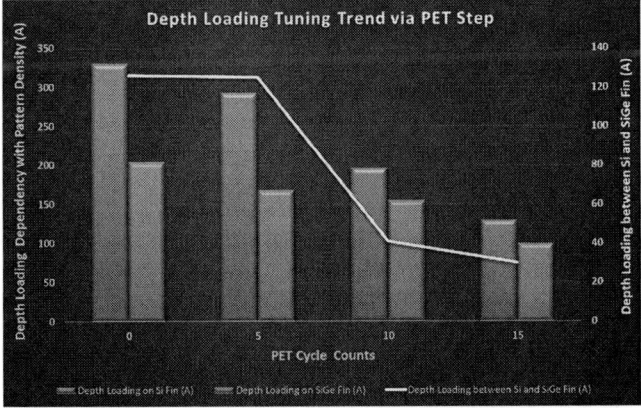

Figure 8. Depth loading performance between Si and SiGe Fin via varying cycles of PET step.

For STI Fin protection liner, ALD conformal SiN layer is chosen, which prevents SiGe Fin CD loss during STI oxide fill process and densification step. Figure 9 shows SiN still presents post STI OX CMP step, and comparable Si and SiGe Fin CD is preserved. High quality SiN liner is attributed to prevention of SiGe oxidation during STI oxide densification step.

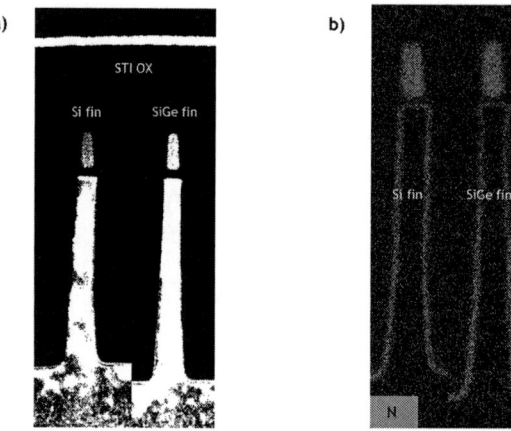

Figure 9. a) Z-contrast STEM of Si (left) and SiGe (right) Fin post STI OX CMP; b) EDX result.

Conclusion

We have developed buried SiGe integration schemes and demonstrated dislocation-free SiGe EPI growth in p-trench patterning area. Less than 40 A of Si and SiGe Fin depth loading is achieved, which is benefited by optimizing repeated cycles of PET step during dual Fin plasma dry etch using advanced pulsing mode. SiN liner provides oxidation free for further STI oxide step, comparable Si and SiGe Fin CD is preserved post STI OX CMP step.

Acknowledgments

This work was partially supported by "Shanghai Pujiang Program".

References

1. D. Guo; G. Karve; G. Tsutsui; K-Y Lim; R. Robison; T. Hook; R. Vega; D. Liu; S. Bedell; S. Mochizuki; F. Lie; K. Akarvar, M. Wang, R. Bao, S. Burns, V. Chan, K. Cheng, J. Demarest, J. Fronheiser, P. Hashemi, J. Kelly, J. Li, N. Loubert, P. Montanini, B. Sahu, M. Sankarapandian, S. Sieg, J. Sporre, J. Strane, R.

Southwick, N. Tripathi, R. Venigalla, J. Wang, K. Watanabe, C. W. Yeung, D. Gupta, B. Doris, N. Felix, A. Jacob, H. Jagannathan, S. Kanakasabapathy, R. Mo, V. Narayanan, D. Sadana, P. Oldigies, J. Stathis, T. Yamashita, V. Paruchuri, M. Colbum, A. knorr, R. Divakaruni, H. Bu, and M. Khare, *VLSI,* (2016).

2. R. Bijesh; I. Ok; M. Baykan; C. Hobbs; P. Majhi; R. Jammy; S. Datta, *69th Device Research Conference,* (2011).

3. M. L. Lee, E. A. Fitzgerald, M. T. Bulsara, M. T. Currie, and A. Lochtefeld, *J. Appl. Phys.*, **97**(1), (2005).

4. Y. Sun, S. E. Thompson, and T. Nishida, *J. Appl. Phys.*, **101**(10), (2007).

5. Y. Suzuki, S. Ogiwara, T. Hosoi, T. Shimura, and H. Watanabe, *Appl. Phys. Lett.,* **101**(20), (2012)

6. Y. Ishii, Y. Lee, W. Wu, K. Maeda, H. Ishimura, and M. Miura, *EDTM,* (2019)

7. T. David, A. Benkouider, J. N. Aqua, M. Cabie, L. Favre, T. Neisius, M. Abbarchi, A. Naffouti, A. Ronda, K. L. Liu, and I. Berbezier, *J. Phys. Chem. C,* **119**(43), (2015)

8. N. R. Zangenberg, J. Lundsgaard Hansen, J. Fage-Pedersen, and A. Nylandsted Larsen, *Phys. Rev. Lett.*, **87**(12), (2001)

9. G. Tsutsui, H. Zhou, A. Greene, R. Robison, J. Yang, J. Li, C. Prindle, J. R. Sporre, E. R. Miler, D. Liu, R. Sporer, B. Mulfinger, T. McArdle, J. Cho, G. Karve, F. L. Lie, S. Kanakasabapathy, R. Carter, D. Gupta, A. Knorr, D. Guo, and H. Bu, *VLSI,* (2017).

ECS Transactions, 104 (4) 217-227 (2021)
10.1149/10404.0217ecst ©The Electrochemical Society

Highly Selective SiGe Dry Etch Process for the Enablement of Stacked Nanosheet Gate-All-Around Transistors

C. Durfee[1], S. Kal[2], S. Pancharatnam[1], M. Bhuiyan[1], I. V. Otto IV[2], M. Flaugh[2], J. Smith[2], D. Chanemougame[2], C. Alix[2], H. Zhou[1], J. Frougier[1], A. Greene[1], M. Belyansky[1], K. Watanabe[1], J. Zhang[1], D. Schmidt[1], M. Breton[1], K. Zhao[1], M. Wang[1], V. Basker[1], A. Mosden[2], N. Loubet[1], D. Guo[1], P. Biolsi[2], B. Haran[1], and H. Bu[1]

[1] IBM Research, 257 Fuller Road, Albany, NY 12203, USA
[2] TEL Technology Center, America, LLC, 255 Fuller Rd., Albany, NY 12203, USA

> Horizontally stacked nanosheet gate-all-around devices enable area scaling of transistor technology, while providing improved electrostatic control over FinFETs for a wide range of channel widths within a single chip for simultaneous low power applications and high-performance computing. Fabrication of inner spacers and Si channels is challenging, but essential to device performance, yield, and reliability. We elucidate these challenges and detail their impact to the device. We overcome these challenges with novel, highly selective, isotropic SiGe dry etch techniques which enable precise, robust inner spacer and channel formation. Finally, we demonstrate substantial improvements to relevant device parameters: resistance, drive current, transconductance, threshold voltage, breakdown voltage, bias temperature instability and overall variability.

Introduction

To continue future device scaling beyond 5nm, horizontally stacked gate-all-around nanosheet (GAA NS) device architecture provides one of the best area scaling options with good electrostatic control [1-5]. A typical NS process integration scheme (Fig. 1) involves several key components that control device performance: the epitaxial SiGe/Si superlattice, which defines the Si channel thickness (T_{Si}) and the sacrificial SiGe thickness (T_{sus}), or "suspension layer"; the inner spacer (IS) module and the high K metal gate (HKMG) module [4-8]. The shape and depth of IS formed by indenting the SiGe layers determines the HKMG placement and gate length (L_g), and is essential for protecting the source-drain (S/D) epitaxy during channel release (CR) [4,9]. High selectivity and carefully tuned SiGe etch rates (ERs) during IS indentation and CR preserve the thickness and shape of the channel, which is critical for good device performance. In this paper, we explain some of the key integration challenges of GAA NS architecture, introduce novel gas phase SiGe etches that are optimized to provide solutions to these challenges, and correlate the impact of these etches on long channel (LC) device electrostatics for a wide range of NS widths (W_{NS}), where $L_g = 100$nm and $W_{NS} = 20$-100nm. This study helps identify the key etch requirements for IS formation and CR to guide future improvements in the NS device performance.

217

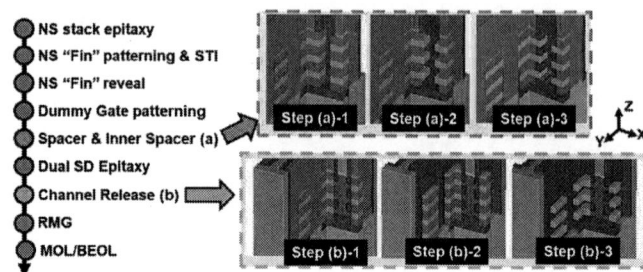

Figure 1. A typical stacked NS GAA integration sequence. (a) IS formation, and (b) CR process steps shown across dummy gate (x direction) and across W_{NS}/"Fin" (y direction).

Nanosheet Architecture Challenges

GAA NS integration presents many challenges that have a critical impact on device performance. Some of these challenges, detailed in Table 1, in part are a result of non-optimal IS and HKMG formation driven by the IS indent and CR etches. For IS indent, the etch process needs to have good control to provide a well-calibrated depth for optimal overlap capacitance (C_{ov}) and R_{on} [9]. The process needs excellent selectivity to prevent Si extension (T_{ext}) thinning, minimizing external resistance (R_{ext}). Etch-front rounding needs to be minimized during the IS SiGe etch to prevent etch paths between the sacrificial SiGe and the S/D epitaxy during CR, which would result in damaged S/D epitaxy and compromising device performance [4,9]. Non-uniform etching along the width of the sheet needs to be reduced to prevent corner weakness, which also can cause S/D epitaxial damage. In addition, etch damage to the spacers and hardmask must be eliminated to avoid non-selective epitaxial growth, S/D epitaxial damage, or damage to the dummy gate, which can impact HKMG formation and device performance. For CR, high selectivity is critical to minimize Si channel surface roughness and T_{Si} loss, which leads to mobility degradation, high channel resistance (R_{ch}) and undesired T_{Si} variability between short channel (SC) and LC devices. The etch process also needs to completely etch the SiGe with small T_{sus} without becoming self-limited to enable a wide range of W_{NS} in LC devices. Each of these challenges is addressed with our highly-selective gas phase etch processes described below.

TABLE I. NS architectural challenges at IS and CR.

Device parameter	Structural parameter
IS formation (SiGe indent; step (a)-2)	
C_{ov}, R_{ext}, R_{on}, DIBL	IS depth
Optimal S/D Epi; Gate metal isolation	IS shape (both across gate and across W_{NS})
R_{ext}, R_{on}	T_{ext} Si loss
CR (Complete SiGe etch; step (b)-3)	
R_{ch}, R_{on}, W_{eff}, gm_{max}, DIT, I_d, BTI	Channel damage (T_{Si} loss, corner rounding, surface roughness)
V_t variability with W_{NS}	Si channel thinning across W_{NS}
Gate conductance, DIT	Residual SiGe

SiGe Dry Etch Process

High selectivity and good ER control are the essential features of the etch processes to address the challenges discussed above. We characterized our SiGe dry etch processes using blanket films, multilayers (ML) and integrated NS structures. Selectivity is determined by SiGe ER, which is a function of Ge% and T_{sus}. SiGe ER increases for low Ge%, reaching a maximum well above the typical Ge% values used for NS architecture (Figs. 2). The ER is linear with higher T_{sus}.

While the highest SiGe ER provides the best selectivity, the optimal Ge% used for the sacrificial SiGe layers is determined by a balance between film strain and selectivity. Strain in the epitaxial SiGe/Si superlattice increases with Ge% and T_{sus}. Above the critical thickness, this strain causes the film to relax, inducing defects in the Si channel and impacting mobility. The selectivities of these processes, determined on actual integrated structures as discussed below, exceed that required to provide excellent IS and Si channel formation while staying well within the critical thickness limit.

Cross-wafer IS etch uniformity is necessary to enable uniform L_g and maximize yield, especially for NS architecture. OCD scatterometry was implemented to provide in-line, high-density non-destructive metrology for IS indent characterization and monitoring. OCD data shows that Process B has 4X better uniformity than that of Process A (Fig. 2 (d)). Since IS formation relies on a timed etch without an etch stop layer, ER control is important for uniformity. ER can be controlled over an order of magnitude without impacting the integrated structure, as shown in (Fig. 2 (a)). While cross wafer uniformity is important for IS indent, it is not as critical at CR, which requires extremely high selectivity with the ability to completely etch SiGe in tight aspect ratios.

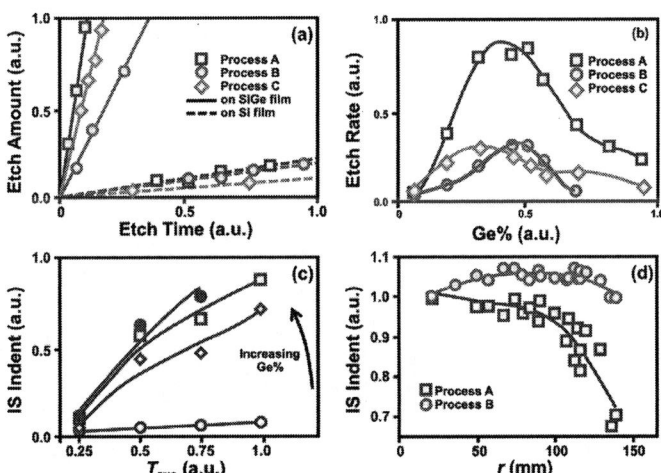

Figure 2. (a) SiGe and Si etch on blanket films. (b) SiGe ER vs. Ge% on MLs. (c) Process A SiGe ER vs. T_{sus} is linear at high T_{sus} for all Ge%. (d) IS etch cross-wafer uniformity by OCD. Process B has 4X better uniformity than Process A.

These processes have highly isotropic selectivity and ERs, as shown by (100) and (110) data (Table 2), making ER anisotropy and subsequent channel corner rounding in NS devices only a function of device geometry and etch selectivity to Si.

Each of these gas phase processes have unique characteristics that make them well suited to meet the IS and CR requirements.

TABLE 2. (Top row) Blanket film SiGe:Si selectivity, and Si etch isotropy.

Structural parameter	Process	
	A	**B**
T_{ext} & T_{si} impact = SiGe (110):Si(100)	43	15
Corner rounding = Si (100):Si(110)	1.1 +/- 0.2	1.0 +/-0.2

Inner Spacer Formation

As discussed above, IS formation determines L_g, HKMG placement, and protects S/D EPI during CR. A robust IS etch process requires highly selective to hard mask, gate spacer, and Si with good process control, sheet-to-sheet variability, and cross-wafer uniformity for a wide range of gate lengths and W_{NS}. It must have minimal profile rounding and a uniform etch front along the sheet width.

Low selectivity can cause T_{ext} thinning; however, both Processes A and B show less than 1nm loss for a 6nm extension indicating > 90% e/g (Fig. 3); e/g = 100% for the ideal case. A slower ER provides better etch control for good top-bottom and left-right IS uniformity; Process B has 12% better SiGe indent uniformity (Fig. 2(d)). The SiGe indent uniformity directly translates to more uniform L_g and HKMG around the Si channel and lead to devices with low variability.

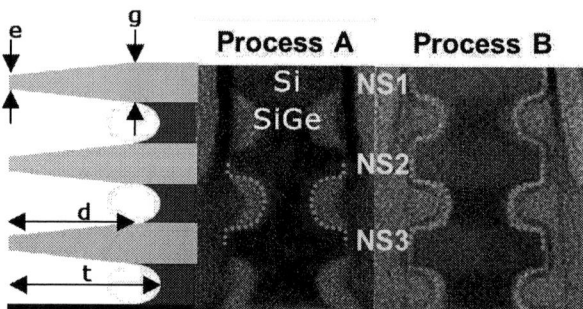

Figure 3. Quantification technique: etch front profile (d/t) and T_{Si} retention (e/g)

TABLE 3. Comparing indent performance between Processes A and B

Structural parameter	Process A	Process B
Si Retention (*e/g*)	96%	93%
IS profile (*d/t*)	61%	82%

Process B has a better IS profile ($d/t \sim 82\%$) than Process A ($d/t \sim 61\%$); $d/t = 100\%$ for the ideal square profile (Fig. 3). We attribute this improvement to the smaller SiGe ER sensitivity to Ge% of Process B. The maximum achievable d/t by IS etch is limited by Ge diffusion at the Si-SiGe interface in the NS superlattice driven by anneals prior to IS etch (Fig. 4) [3-5,9]. For Process B, the IS etch profile ($d/t \sim 82\%$) matches closely with the simulated diffusion profile, indicating a highly optimized IS etch process. In addition, removing the diffused SiGe layer during the indent etch and preventing seam formation during IS atomic layer deposition (ALD) are important to avoid etch paths between the SiGe layer and the S/D epitaxy during CR, which would damage the S/D epitaxy [9].

As discussed previously, a uniform SiGe etch front profile along W_{NS} prevents additional pathways for S/D damage during downstream processes [9, 10]. The plan view TEM (Fig. 5) shows a straight SiGe etch front for even the widest W_{NS} of 100 nm.

Although Process A has better selectivity, Process B is the best process for IS indent; its selectivity is sufficient to deliver good T_{ext} while providing a better etch front for S/D protection during CR.

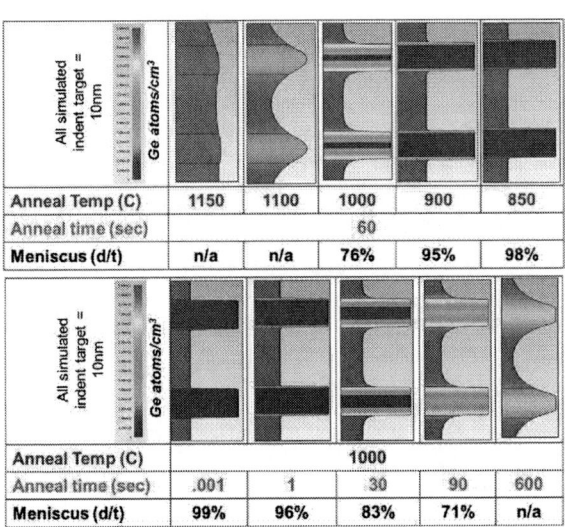

Figure 4. Effect of anneal temperature (*top*) and duration (*bottom*) on IS profile (*d/t*) for Process A. Modeling based on Fig. 2(b) and Reference [4].

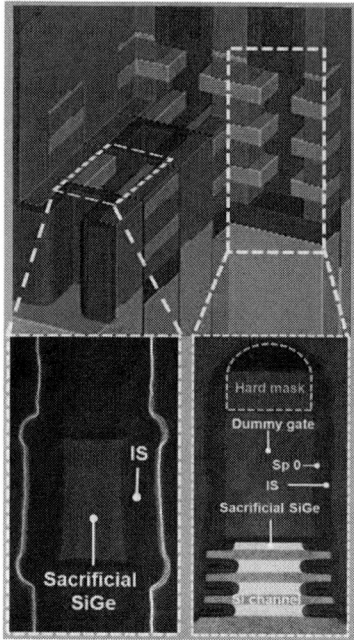

Figure 5. (*Top*) IS schematics: (*Left*) profile along W_{NS}. (*Right*) profile across gate.

Channel Release

A highly-selective CR process is critical for formation of pristine Si channels and subsequent HKMG. Selectivity to Si and dielectrics enables complete removal of all residual SiGe at the Si-SiGe interface without impacting the Si channel, plus gives smooth channel surfaces with no corner rounding, providing uniform Si channels with maximum effective sheet width (W_{eff}), which enables optimal electrical performance. Also, the process must not be self-limited for small T_{sus} to enable uniform HKMG formation across a wide range of W_{NS} for both SC and LC devices for power-performance optimization. Insufficient selectivity leads to T_{Si} loss or residual Ge at the Si-SiGe interface, Si surface roughness and channel corner rounding [7,9], leading to mobility degradation, high channel resistance (R_{ch}) and T_{Si} variability.

We have developed three SiGe etch processes with selectivities ranging between 100:1 to greater than 800:1, with peak ERs at about 30% SiGe – a typical value for the NS structures – and Si channel RMS roughness ~0.3Å (Fig. 2 (b)) [9]. As a result of higher SiGe etch selectivity, Processes A and C result in minimal corner rounding and T_{Si} loss compared to Process B (Fig. 6). Process A exhibited no damage above 2nm IS thickness (Fig. 7); indicating low permeability of the etch chemistry through the IS during CR, which is necessary to prevent S/D damage [4, 9, 11]. Process A at CR enables excellent structural integrity of the device post CR and post the first metal layer (M1) with intact dummy gate spacer, IS, S/D epitaxy and well-formed HKMG (Fig. 8).

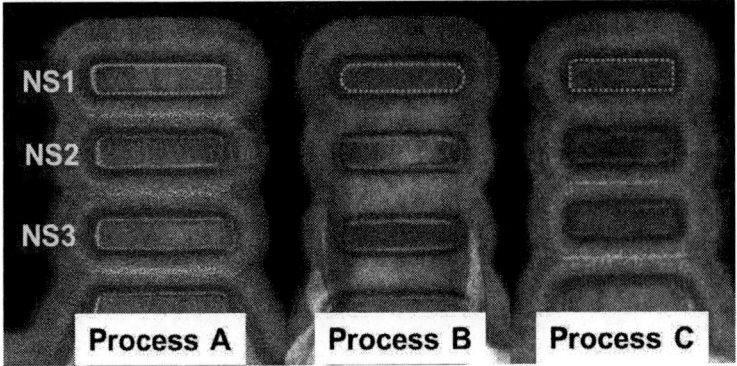

Figure 6. Si channel comparison of Processes A, B and C at CR. TEMs obtained post M1 test.

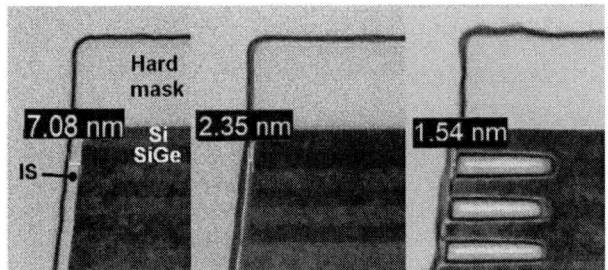

Figure 7. Process A etch gas only penetrates IS thicknesses below 2nm on Si-SiGe ML test structures encapsulated with an ALD IS film.

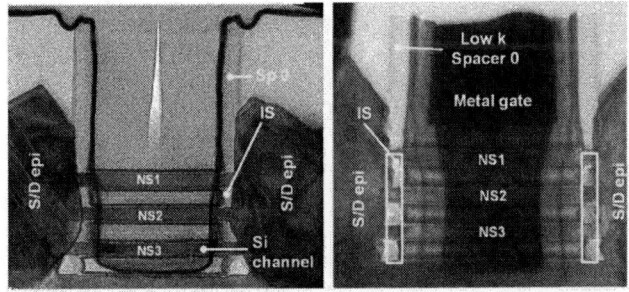

Figure 8. Si channel and S/D epitaxial structural quality. Cross-gate TEM post CR (*left*) and post HKMG (*right*) shows well-preserved S/D epitaxy, uniform IS and intact gate spacer (or spacer 0).

Ge diffusion limits the ability to form a perfect HKMG structure. Theoretical simulations for various anneals (Fig. 4), predict 0.7-1.2 nm Ge diffusion during our process sequence, which is consistent with HRTEM (Fig. 9) of about 1-2nm [4]. This diffused layer is completely etched during CR by both processes, reducing T_{Si} but leaving a pristine Si channel as confirmed by Electron Dispersive X-ray Spectroscopy (EDX) (Fig. 10). Complete Ge removal is important to reduce interfacial trap density (DIT), improving gate conductance [11].

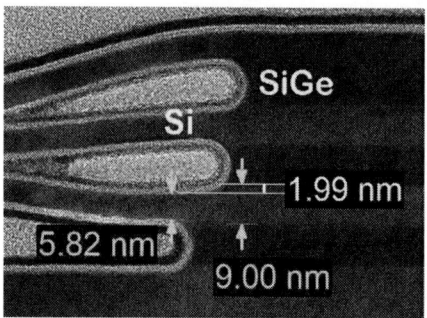

Figure 9. HRTEM on 100nm SiGe etched ML test structure shows 2nm Si loss due to Ge diffusion, consistent with Fig. 4 modeling and Fig. 10 EDX.

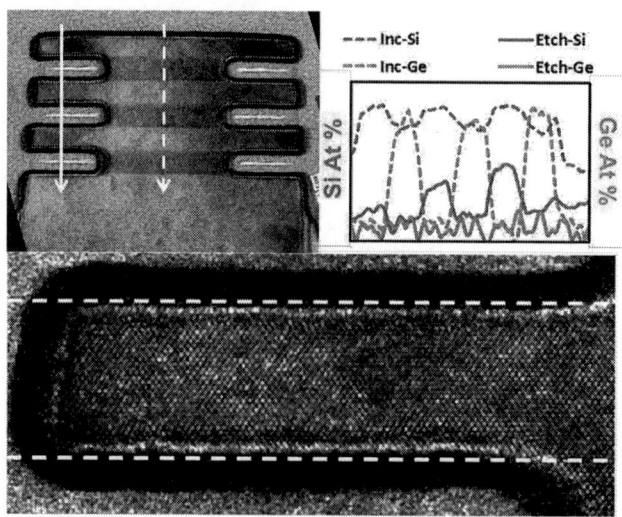

Figure 10. HRTEM on 100nm SiGe-etched ML test structure. EDX line scan shows complete SiGe removal. HRTEM reveals excellent selectivity and atomically smooth surfaces [9].

Electrical Results

Processes A and C show improved device performance over Process B at CR for drive current (I_d) and transconductance (gm). nFET R_{on} for Processes A and C is ~10% better than Process B for W_{NS} = 20nm (Fig. 11(a)), which translates to ~10% improvement in R_{ch} (Fig. 11 (b)), driven by a similar improvement in peak transconductance (gm_{max}) (Fig. 11 (d)). The R_{ext} impact from all 3 processes at CR is negligible due to minimal T_{Si} loss at IS formation (Fig. 11(c)), consistent with e/g in Fig. 3, table 3. This improvement in R_{on} is primarily driven by minimal surface roughness, reduced T_{Si} loss and less corner rounding of the Si channel from Processes A and C (Fig. 6), resulting in larger device W_{eff}, which in turn results in higher gate capacitance (Fig. 12). Previously, we showed sub-monolayer surface roughness impact with 50X overetch from the dry SiGe etch [9]. A larger W_{eff} from Process A and C over B is further validated by the 40mV reduction in threshold voltage (V_t) (Fig 13) and additional V_t degradation resulting from increased Process B overetch. Therefore, we can attribute the primary component of the improvement in R_{on} to R_{ch}, which results from higher W_{eff} with Processes A and C at CR. Correspondingly, breakdown voltage (VBD) and bias temperature instability (BTI) also improve with Processes A and C, with reduced variability in BTI (Fig. 14).

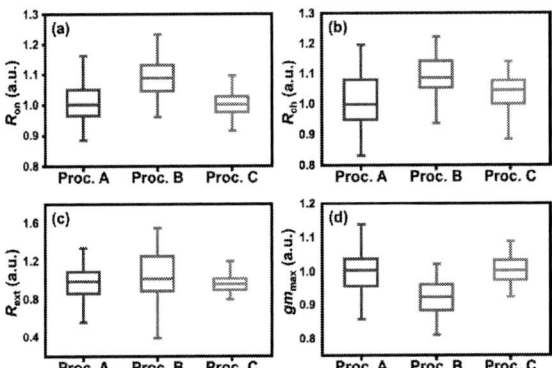

Figure 11. (a) gm_{max}, (b) R_{on}, (c) R_{ch}, (d) R_{ext} for Processes A and B at CR for W_{NS} = 20nm and L_g = 100nm. Process B was used for IS indent in both cases.

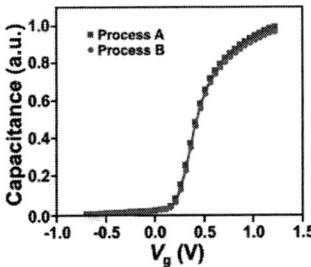

Figure 12. Capacitance vs. V_g with Processes A and B for W_{NS} = 20nm and L_g = 100nm.

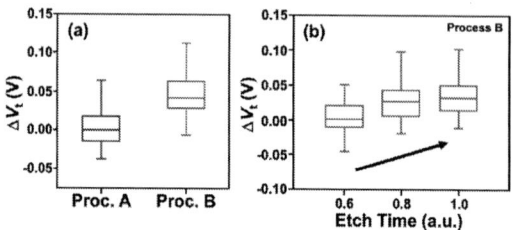

Figure 13. (a) V_t impact of Processes A and B at CR for W_{NS} = 20nm and L_g = 100nm. (b) V_t degradation from overetch with Process B.

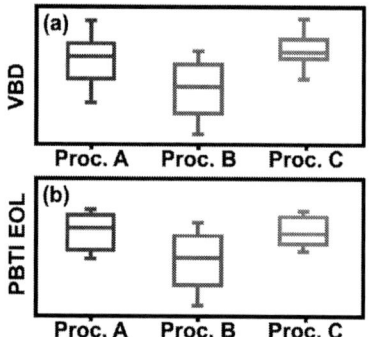

Figure 14. VBD and PBTI are better with Process A and C at CR for W_{NS} = 20nm and L_g = 100nm. Process B was used at IS indent in both cases.

Conclusions

NS device architecture presents many challenges during IS and HKMG formation that strongly impact device performance. We have developed three highly uniform SiGe dry etch processes that significantly improve stacked NS GAA device performance. These processes are specifically optimized to maximize device performance by implementing Processes A and C at CR and Process B at IS indent. The high selectivity of Processes A and C, coupled with its ability to etch tight aspect ratios addresses each of the challenges at CR by creating uniform Si channels across the wafer and across all sheet widths that are devoid of surface roughness and residual SiGe. Conversely, Process B delivers a well-controlled etch with optimal IS shape and cross-wafer uniformity at IS indent, enabling structurally robust devices with better preserved spacers, hardmask and S/D epitaxy. These processes address the challenges discussed previously, maximizing device performance. The high structural integrity provided by these processes enables improved W_{eff}, enhanced gm_{max}, higher I_d, reduced variability in L_g, lower R_{on} and constant V_t across W_{NS}, facilitating higher yield for scaling of future logic technology nodes.

References

1. K. J. Kuhn, *Trans. Electron Devices*, **59**, 7, (2012).
2. D. Nagy, G. Indalecio, A. J. GarcíA-Loureiro, M. A. Elmessary, K. Kalna, and N. Seoane, *IEEE J. Electron Devices Soc*, **6**, 332, (2018).
3. H. Mertens, R. Ritzenthaler, A. Chasin, T. Schram, E. Kunnen, A. Hikavyy, L. Ragnarsson, H. Dekkers, T. Hopf, K. Wostyn, K. Devriendt, S. Chew, M. Kim, Y. Kikuchi, E. Rosseel, G. Mannaert, S. Kubicek, S. Demuynck, A. Dangol, N. Bosman, J. Geypen, P. Carolan, H. Bender, K. Barla, N. Horiguchi and D. Mocuta, *IEDM*, 524 (2016).
4. N. Loubet, T. Hook, P. Montanini, C-W. Yeung, S. Kanakasabapathy, M. Guillom, T. Yamashita, J. Zhang, X. Miao, J. Wang, A. Young, R. Chao, M. Kang, Z. Liu, S. Fan, B. Hamieh, S. Sieg, Y. Mignot, W. Xu, S-C. Seo, J. Yoo, S. Mochizuki, M. Sankarapandian, O. Kwon, A. Carr, A. Greene, Y. Park, J. Frougier, R. Galatage, R. Bao, J. Shearer, R. Conti, H. Song, D. Lee, D. Kong, Y. Xu, A. Arceo, Z. Bi, P. Xu, R. Muthinti, J. Li, R. Wong, D. Brown, P. Oldiges, R. Robison, J. Arnold, N. Felix, S. Skordas, J. Gaudiello, T. Standaert, H. Jagannathan, D. Corliss, M-H. Na, A. Knorr, T. Wu, D. Gupta, S. Lian, R. Divakaruni, T. Gow, C. Labelle, S. Lee, V. Paruchuri, H. Bu, and M Khare, *VLSI*, T230 (2017).
5. N. Loubet, J. Li, R. Chao, C. Yeung, J. Frougier, C. Durfee, A. Arceo de la Pena, R. Muthinti, Z. Bi, M. Sankarapandian, W. Xu, Y. Mignot, S. Sieg, R. Conti, B. Veeraraghavan, H. Jagannathan, B. Haran, R. Divakaruni, and H. Bu, *ECS Meeting Abstracts*, **31**, 1075, (2018).
6. C. W. Yeung, J. Zhang, R. Chao, O. Kwon, R. Vega, G. Tsutsui, X. Miao, C. Zhang, C-W. Sohn, B. K. Moon, A. Razavieh, J. Frougier, A. Greene, R. Galatage, J. Li, M. Wang, N. Loubet, R. Robison, V. Basker, T. Yamashita, and D. Guo, *IEDM*, 652 (2018).
7. M. Wang, J. Zhang, H. Zhou, R. Southwick, R.H Kuo Chao, M. Xin, V. Basker, T. Yamashita, D. Guo, G. Karve, H. Bu, and S. James, IRPS, **1**, 6, (2019).
8. R. Ritzenthaler, H. Mertens, V. Pena, G. Santoro, A. Chasin, K. Kenis, K. Devriendt, G. Mannaert, H. Dekkers, A. Dangol, Y. Lin, S. Sun, Z. Chen, M. Kim, J. Machillot, J. Mitard, N. Yoshida, N. Kim, D. Mocuta, and N Horiguchi, IEDM, 508, (2018).
9. N. Loubet, S. Kal, C. Alix, S. Pancharatnam, H. Zhou, C. Durfee, M. Belyansky, N. Haller, K. Watanabe, T. Devarajan, J. Zhang, X. Miao, M. Sankar, M. Breton, R. Chao, A. Greene, L. Yu, J. Frougier, D. Chanemougame, K. Tapily, J. Smith, V. Basker, A. Mosden, P. Biolsi, T. Hurd, R. Divakaruni, B. Haran, and H. Bu, IEDM, 242, (2019)
10. N. Loubet, T. Kormann, G. Chabanne, S. Denorme, and D. Dutartre, Thin Solid Films, **517 (1)**, 93-97 (2008).
11. G. Bae, D. Bae, M. Kang, S. Hwang, S. Kim, B. Seo, T. Kwon, T. Lee, C. Moon, Y. Choi, K. Oikawa, S. Masuoka, K. Chun, S. Park, H. Shin, J. Kim, K. Bhuwalka, D. Kim, W. Kim, J. Yoo, J. Jeon, M. Yang, S. Chung, B. Ham, K. Park, W. Kim, G. Song, Y. Kim, M. Kang, K. Hwang, C. Park, J. Lee, S. Jung, and H. Kang, IEDM, 287, (2018).

Author Index

Alix, C.	217	Hikavyy, A. Y.	139, 167
Arita, M.	93	Hirose, M.	113
Ayyad, M.	139, 167	Hizawa, T.	147
		Horiguchi, N.	139
Basker, V.	217		
Beam, E.	51	Ikeda, N.	121
Belyansky, M.	217	Inoue, M.	113, 121, 129
Bhuiyan, M.	217	Irokawa, Y.	113
Biolsi, P.	217	Ishihara, N.	83
Breton, M.	217	Ishikawa, Y.	147
Briggs, B.	139	Ito, H.	83
Bu, H.	217		
		Kal, S.	217
Cao, Y.	51	Katamawari, R.	147
Chanemougame, D.	217	Katase, T.	93
Cheng, K.	51	Kawashita, K.	147
Chida, K.	33	Ke, X.	201, 209
Claeys, C.	3	Kita, K.	193
		Koide, Y.	113
Deng, W. F.	209	Kurosawa, M.	183
Du, Z.	51		
Durfee, C.	217	Langer, R.	139
		Lee, T. E.	17
Favia, P.	139	Loo, R.	139, 167
Flaugh, M.	217	Loubet, N.	217
Frougier, J.	217	Luo, X.	17
Fujiwara, A.	33		
		Ma, Y.	51
Greene, A.	217	Machida, K.	83
Guo, D.	217	Maeda, E.	113
		Makihara, K.	105
Hanafusa, H.	63	Masu, K.	83
Haran, B.	217	Matagne, P.	3
Hashizume, T.	113	Matsuguchi, K.	63
Higashi, S.	63	Mencarelli, M.	139

Miyake, Y.	83	Smith, J.	217
Miyazaki, S.	105	Sonoi, S.	147
Morita, Y.	129	Su, B.	201, 209
Morris, R. J. H.	167		
Mosden, A.	217	Tachibana, M.	147
Motomura, K.	147	Tahara, K.	17
		Takagi, S.	17
Nabatame, T.	113, 121, 129	Takahashi, Y.	93
Nagase, M.	27	Takenaka, M.	17
Nakane, R.	17	Toprasertpong, K.	17
Nakashima, H.	157	Tsukagoshi, K.	113, 121
Nakatsuka, O.	183	Tsurumaki-Fukuchi, A.	93
Nako, E.	17		
Nishiguchi, K.	33	Vantomme, A.	167
Noguchi, K.	147	Veloso, A.	3
Ochi, R.	113	Wang, D.	157
Oh, H.	209	Wang, H.	51
Ohi, A.	121	Wang, M.	217
Ohta, H.	93	Wang, Z.	17
Onaya, T.	113, 121, 129	Watanabe, K.	217
Ota, H.	129		
Otto IV, I.	217	Xiao, M.	51
Oyamada, R.	147	Xiao, X.	201
		Xie, A.	51
Pancharatnam, S.	217		
Porret, C.	139, 167	Yamamoto, K.	157
Pourtois, G.	167	Ye, B.	209
		Yin, C.	209
Rengo, G.	167		
Rosseel, E.	167	Zhang, E. N.	209
		Zhang, H. Y.	201, 209
Sato, T.	63	Zhang, J.	217
Sawada, T.	113, 121, 129	Zhang, Y.	51
Schmidt, D.	217	Zhao, J.	209
Shiojima, K.	69	Zhao, K.	217
Shiozaki, K.	113	Zhou, H.	217
Simoen, E.	3		